591.57
C625t

Tooth and Claw

J. L. Cloudsley-Thompson

Tooth and Claw

Defensive Strategies in the Animal World

J. M. Dent & Sons Ltd
London   Melbourne   Toronto

First published 1980

© John Cloudsley-Thompson, 1980

All rights reserved. No part of this publication may be reproduced, stored in a retrieval system, or transmitted, in any form or by any means, electronic, mechanical, photocopying, recording or otherwise, without the prior permission of J. M. Dent & Sons Ltd

Designed by Malcolm Young
Phototypeset in 11/13 pt VIP Bembo by Trident Graphics Ltd, Reigate, Surrey
Printed in Great Britain by Billing & Sons Ltd, London, Guildford, Oxford, Worcester
for J. M Dent & Sons Ltd
Aldine House, Welbeck Street, London

British Library Cataloguing in Publication Data

Cloudsley-Thompson, John Leonard
   Tooth and claw.
   1. Animal defences
   I. Title
   591.5'7   QL759

ISBN 0–460–04360–9

Contents

List of Illustrations   6
Introduction   11

1   Life in a Hole   17
2   Camouflage   28
3   Disguise   40
4   Armour   56
5   Barbs and Spines   71
6   Chemical Defences   84
7   Venoms   101
8   Warning and Threat   128
9   Mimicry   140
10   Bluff, Death Feigning and Deflection of Attack   157
11   Withdrawal, Escape and Flight   174
12   Horns, Teeth and Claws   190
13   Co-operation   201
14   Predator-Prey Interactions   220

Bibliography   233
Index   243

List of Illustrations

1   Echiuroid worm in its burrow with associated commensals *19*
2   'Pancake-tortoise' *20*
3   Gribbles *22*
4   Examples of phragmosis *24*
5   Ptarmigan: summer and winter *30*
6   Purple sea-snail *34*
7   Leaf-insect and stick-insect *42*
8   *Cossyphus* *43*
9   Pupae of two species of moth *(Spalgis)* *49*
10 Alligator-bug and Caiman *50*
11 Hunting spider *57*
12 Coffer-fish and pine-cone fish *65*
13 *Triceratops* *67*
14 Pangolins *69*
15 Gin-traps of beetle pupa *(Alphitobius)* *74*
16 Darkling beetle *85*
17 Skunks *95*
18 Scorpion, showing pectines *109*
19 Tarantula and 'malmignatte' *111*
20 Scolopendra *113*
21 Assassin-bug *114*
22 Sea-wasp *117*
23 Stinging cells of a sea-anemone *118*
24 Venom apparatus of marine cone-shell *120*
25 Sea-slug *131*
26 Death's-head hawk-moth *134*
27 Porcupine *138*
28 Spider carrying dead ant *142*
29 Spiders mimicking ants *144*

30 Aardwolf and hyena *146*
31 Indian cobra *160*
32 American opossum *163*
33 Fairy armadillo *175*
34 Flying fish, flying gurnard, flying frog, flying dragon, flying phalanger and flying squirrel *187*
35 Horns and teeth in Africa *191*
36 Variety of horns *197*
37 Sponge-crab *213*
38 Meerkats *221*

Acknowledgment

As usual, my warmest thanks to my wife, Anne Cloudsley, for her shrewd comments – and for drawing the *Triceratops*, ptarmigans and mammals.

In loving memory of BARBIE
1890–1979

Introduction

Debate as to whether aggressive behaviour in Man is instinctive or learned has directed much popular attention to the biology of offence, but the equally fascinating defensive mechanisms of animals, and defence is after all the reciprocal of attack, have been largely neglected. It is, of course, true that in human warfare offence will, in the long run, always overcome defence. Every fortress can eventually be breached, and a combination of mobility and powerful offensive weapons is usually invincible. Nevertheless, defence cannot be discounted in the strategy for survival. Despite the fact that the arrows of English archers at the battles of Crécy, Poitiers and Agincourt in the fourteenth and fifteenth centuries proved superior to the armour of French chivalry and announced to Europe that heavy cavalry had eventually been superseded, this was only after a thousand years of the supremacy of armour on the battlefields of the world. Again, although tanks have subsequently proved vulnerable to anti-tank guns, they were almost invincible at Cambrai in November 1917.

Defensive strategies are, too, at times complementary to, and closely linked with, offence. Even the heaviest ground weapons need to be camouflaged for defence from aerial attack, and armour cannot yet be dispensed with. Since civilization's earliest days, military commanders have been faced with the problems caused by the conflicting requirements of movement, striking power and security. In the animal world, also, there are conflicting requirements. Armour and weapons hamper mobility even though they enhance security. Against some predatory animals it is better to stand and fight: against others, however, flight may be the only policy giving any chance of survival. An animal invulnerable to one type of enemy is often susceptible to attack by another. For instance, the king cobra or hamadryad is the most

Introduction

dangerous of poisonous snakes on account of its great size, aggressive disposition, powerful fangs, and proportionately large amount of lethal poison. It has been known to kill an elephant, which is quite capable of defending itself against mammalian carnivores as powerful as the tiger. On the other hand, even quite large cobras may themselves be vulnerable to the small mongoose. Mongooses do have some measure of resistance to cobra venom, but they rely mainly on agility and on their long fur which, when fluffed up, gives them some protection against a bite.

If attack is the first principle of warfare, defence is certainly the second. Whereas attack is mobile, however, defence is essentially static. Similar principles apply in the animal world, and, in both, economy is invariably one of the most important considerations. During World War II, the British produced a large self-propelled gun, the 'Tortoise', with armour up to 230 mm thick; America turned its attention to the construction of an even larger self-propelled gun with frontal armour of 305 mm; while the largest tank ever designed, the German 'Maus', was armed with an enormous 150 mm cannon. Because its weight was over 188 tonnes, the 1,200 h.p. engine could drive it no faster than 20 k.p.h. (12 m.p.h.). None of these gigantic vehicles ever saw active service, and they were all hopelessly ineffectual in relation to their cost. The same kind of considerations apply in the animal world. For example, it is by no means impossible to conceive of a real tortoise having a shell so thick that it would be invulnerable to all its enemies. (Indeed, this may actually be true of adult giant tortoises of the Seychelles, and Galapagos Islands – after reaching full size, they succumb only to climate, accident, disease, or old age.) If the success of a tortoise is to be evaluated in terms of passing on genetic material to its descendants, however, it may be more effective for it to produce a larger number of young which have thinner shells, but which will reach maturity more rapidly and with less expenditure of metabolic energy on their armour.

The defence mechanisms of animals, like all other hereditary characters, have, of course, developed during their evolution – through the agency of natural selection. In some environments, such as tropical islands, the development of formidable armour may be the best strategy for survival. In others, however, a gene

Introduction

for the inheritance of rapid growth at the expense of a thinner shell may, on average, be more successful. In nature, a balanced condition between opposing tendencies is normally maintained. Natural selection can be defined as the survival of the most fit, with the inheritance of those adaptations wherein the fitness lies. It is a continuous process of trial and error, involving all living matter. Successful individuals, those which survive long enough to reproduce, are able to bequeath genetic material to their offspring and subsequent generations. The genes of the unsuccessful are eliminated from the breeding pool of the species.

No hereditary character persists indefinitely in the absence of positive selection. For instance, after many generations of troglodytic existence, cave animals lose their pigmentation and become eyeless. The eyes are delicate structures, and easily damaged, so that an animal living in absolute darkness might be better off without them; but there is clearly no direct advantage in becoming white. Nevertheless, pigments have to be synthesized by the expenditure of metabolic energy. In darkness, where they are not seen, this expense cannot be justified. There is, therefore, a selective advantage to cave animals in becoming colourless. In a similar way, it is not economical, in the absence of predatory enemies, for an animal to retain defensive structures, or even the instinct to escape. This may explain the suicidal tameness and defencelessness of the dodo, the solitaire and the giant Mauritian parrot, all of which lost their powers of flight. Because they were so easily killed by Man, as well as by his pigs and dogs, they became extinct within a few years of the discovery of their tranquil island homes.

One of the most impressive qualities of nature is the teeming abundance of living creatures – the immense numbers of individuals in a great school of fishes, the myriads of wildebeest that migrate annually across the plains of Serengeti, the vast assemblage of micro-organisms in a drop of pond water, or the huge population of human beings inhabiting a large city. The productivity of all living organisms is far beyond what is required for their own replacement. Even at the lowest rate of reproduction, numbers increase in geometric ratio; whereas the space they occupy, and the available food supply, remain relatively constant. The ling, a relative of the cod, may lay nearly 30 million eggs but, as the ultimate number of fishes remains approximately the

Introduction

same, not more than two of these can survive to reproduce. Clearly, only the most healthy and fortunate fish has even the remotest chance of reproducing.

The elephant is the slowest-breeding of all animals. Nevertheless, Charles Darwin calculated that, if all the offspring were to reproduce, the progeny of a single pair of elephants would, after 750 years, number no less than 19 millions! From this, he argued there must be intense selection for hereditary characters that are beneficial, and that only the fittest of individuals will survive. Likewise, T. H. Huxley estimated that the descendants of a single greenfly, if all were to survive and multiply, would, by the end of one summer, weigh down the entire human population of China. An oyster produces up to 60,000,000 eggs: if all the offspring of one oyster were to survive and multiply until there were great-great-grandchildren, these might number some 66,000,000,000,000,000,000,000,000,000,000, and the heap of shells would be eight times the size of the earth. Again, if the present rate of increase of mankind were to continue, within less than 1,000 years a mass of people would be standing on each others' shoulders, more than a million deep. By 2,000 years, the mountain of humanity, travelling onwards at the speed of light, would have reached the edge of the known universe!

With such extraordinary productivity on the part of all living organisms, the efficiency of the check upon the increase of every species of plant or animal becomes immediately apparent. This is why Charles Darwin and Alfred Russel Wallace both recognized as inevitable the 'struggle for existence', the competition between all organisms, and between each individual and the physical environment. This struggle is threefold: environmental, against heat, cold, drought, excess moisture, lightning and tempest, earthquake and volcanic eruption; intraspecific, between members of the same species, for food, territory and mates, or to avoid cannibalism; and interspecific, between members of different species for food and living space and when one species is parasitized or preyed on and eaten by another. As in human warfare, the struggle is often more bitter the closer the relationship or similarity between the contestants.

We are here concerned mainly with defences against predation; but defence against attack by other members of the same species is also an important factor in natural selection, and often the same

Introduction

weapons are used in both contexts. Predation has always played a leading part in the struggle for existence, and is an important cause of death among the majority of animals so that, in consequence, extremely efficient defence mechanisms have been selected since earliest times. (A large proportion of the examples cited refer to the world of insects: this is because insects are more diverse and numerous than any other animals – so, naturally, they provide the greatest variety of adaptations.) During recent years, much attention has been devoted to intraspecific aggression as well as to interspecific predation – somewhat to the neglect of defence. These are, however, opposite aspects of the same phenomenon; and one of the objects of the present work will be to help redress the balance.

1 Life in a Hole

In modern warfare, the natural reaction of the infantryman, whenever possible, is to dig himself a fox-hole or slit trench in which he can obtain protection from the blasts of exploding shells and lie concealed from the unwelcome attentions of the enemy. Against troops in the open, well-directed bursts of machine-gun fire and high explosive may be unequalled killers but, against men in trenches, they are practically impotent. At the same time, however, no military engagement can be brought to a satisfactory conclusion by passive defence alone, and some defences must be active. Likewise, the defensive strategies of animals, by which protection from predatory enemies is obtained, may be either passive, active, or more commonly, a combination of both. Primary defences are defined as those which operate regardless of whether a predator is in the vicinity or not: they reduce the chance of an encounter between predator and prey, and consist of hole-dwelling, protective coloration, and the avoidance of detection by sound or smell. The possession of armour is also a primary defence. In contrast, active or secondary defences, such as flight, death feigning, deflection of attack, bluff, defensive associations and retaliation are invoked only after a predator has been detected. Most animals have evolved more than one kind of defence mechanism, and alternative strategies are employed in different circumstances.

Many otherwise defenceless animals spend almost their entire lives hidden from predators in crevices or holes in the ground. Such recluses are sometimes known as 'anachoretes' or 'anchorites' – from a Greek word meaning 'one who has withdrawn or secluded himself from the world'. They include earthworms, burrowing insects and lizards, subterranean snakes, and moles, which scarcely ever emerge from the security of their retreats. In the oceans, too, many worms, molluscs and crustaceans find safety by burrowing into the sand or mud of the sea-bed.

Below the sea-bed, in the earth, or secreted beneath leaf litter and the bark of trees are millions of tiny creatures to which the name 'cryptozoa', meaning hidden life, is applied. The distinction between anchorites and cryptozoa is that the former live a solitary existence whereas cryptozoic animals are usually present in very large numbers. For instance, if you turn over some of the leaf litter in any beech wood, you will immediately find an astonishing number of tiny wingless springtails hopping madly about, as well as microscopic mites crawling among the rotting vegetation, nematode worms, snails, woodlice, centipedes, millipedes, small insects, false-scorpions and spiders. These are the cryptozoa, most of which are herbivores; but some are carnivores which prey on the others. Both groups are hidden from larger predators in the outside world. In fact, most of the animals inhabiting the earth lead hidden lives for part, or all of the time. The creatures we see around us – dogs, cats, rabbits, squirrels, cows, horses, elephants, zebras, birds, butterflies and fishes – represent only a familiar minority.

Many anchorites and cryptozoic animals live almost permanently in their retreats. Others, mainly cryptozoa such as woodlice, centipedes, millipedes, scorpions, spiders and ground beetles, remain in hiding during the day but emerge at night to mate and search for food. These often have poorly developed eyes for perceiving predators by day, but are endowed with well-developed senses for detecting nocturnal enemies.

The activities of many animals are timed to avoid exposure to predation. Small mammals, such as mice, jerboas, bats, bushbabies and rabbits, remain hidden in their holes during the day, when predators could easily see them, but forage at night when they are less readily detected. In contrast, many small birds are active during the day, when their good eyesight and their speed enable them to elude their enemies. At night, however, they roost in trees and holes where they are protected from nocturnal carnivores.

Many marine worms, crustaceans and molluscs, are anchorites and live their entire lives buried in the sand or mud of the seashore. Most of them inhabit naturally occurring crevices or dig their own burrows, but some share the holes and tubes made by

Life in a Hole

animals of other species. For instance, the burrow of an echiuroid worm often contains a scale-worm, a bivalve mollusc, a small crab and a goby-fish. Such an association between individuals of two or more different species, from which both or nearly all benefit, is known as 'commensalism'. Several other examples will be cited in this book. Many marine commensals not only derive

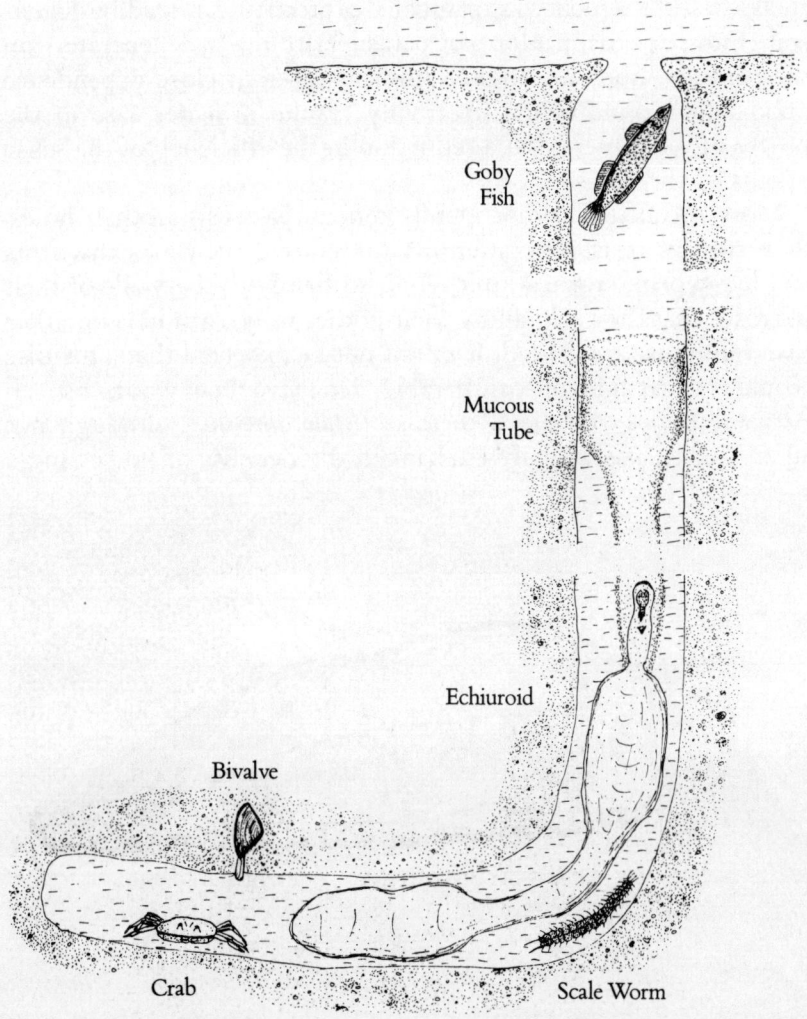

1. Echiuroid worm in its burrow with associated commensals all of which derive protection from predators on the sea-bed by sharing its burrow.

Tooth & Claw

protection from their association, but also obtain food from the currents of water conveniently produced by their hosts. Tiny crabs, for instance, are sometimes found in the tubes of fanworms or within the shells of mussels, oysters and other bivalve molluscs. Usually they do no harm to their hosts. Occasionally, however, adult pea-crabs not only snatch some of the food laboriously filtered from the surrounding water by the gills of the bivalve host, but even damage and eat the gills as well. Thus, an association which was originally protective can readily degenerate into parasitism. Not that parasites are really degenerate – on the contrary, they are highly specialized – but close dependence upon another individual invariably results in a decrease in the power of movement and a reduction in the efficiency of the sense organs of the parasite.

Most anchorites do not merely remain passively in their holes; they stoutly resist any attempts to remove them. Earthworms and lug-worms, for example, cling so firmly to the walls of their burrows that they will allow their bodies to be torn in two rather than relinquish their hold. It might not be expected that a tortoise would be capable of comparable defensive behaviour, but an African species, Tornier's tortoise *(Malacochersus)*, although not an anchorite, has actually exchanged the benefits of possessing a

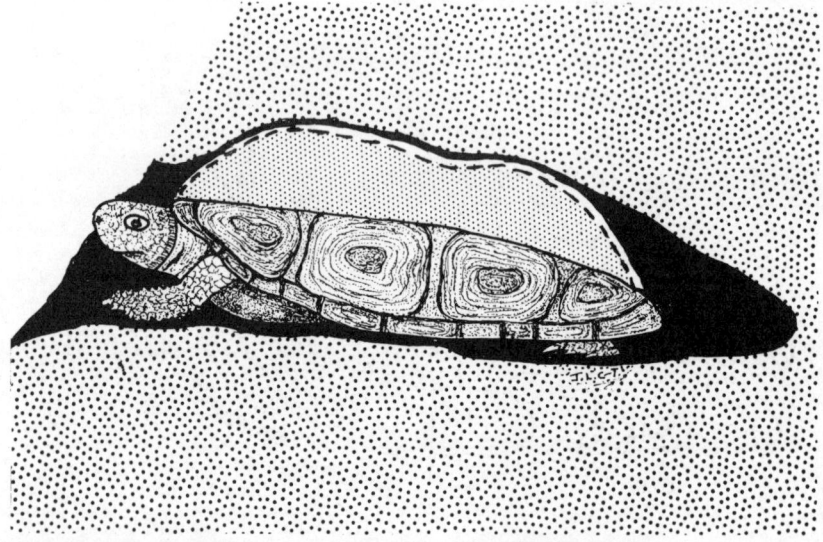

2. How the 'pancake-tortoise' blows itself up to escape capture.

Life in a Hole

hard shell for the advantages of being able to wedge itself into rock crevices. This weird 'pancake-tortoise' is so soft that it can be squeezed out of shape between finger and thumb! It inhabits rugged country and, when threatened, forces itself into narrow clefts between rocks. Then it puffs itself up with air so that it cannot be pulled out. In the environment inhabited by Tornier's tortoise, this ability must be of greater value than would be the possession of a hard shell, so useful in more open country. The efficiency of any defence mechanism depends not only upon the kind of enemy against which it is directed, but also upon the conditions under which its possessor is exposed to attack.

Burrowing animals are usually highly adapted to their modes of life. Their bodies tend to be as nearly cylindrical as the ancestral anatomy of their possessors will permit, and they are almost always pointed in front. Eyes are reduced in size, or even lost, as are ear-flaps and other projections that might hamper movement through the soil. Animals that burrow generally do so either by simply thrusting the ends of their bodies into the substratum, as earthworms do, or by inserting special digging organs, such as the foot of the razor-shell and other molluscs, the proboscis of the acorn-worm *(Balanoglossus)*, or the pointed aboral end of the body of the burrowing sea-anemone *(Edwardsia)*. Earthworms sometimes eat their way through soil, swallowing it if it is too compact for them to force a passage through. When limbs are used for digging, as in the case of the mole or the mole-cricket, they are often strengthened and broadened for the purpose.

Animals that burrow into wood generally do so by means of the mouthparts: for instance, wood-boring beetles, larvae of the goat-moth, and marine, wood-boring crustaceans such as the gribble *(Limnoria)* use their mandibles. The gribble is not completely solitary, like the majority of anchorites, for there is usually a pair to each burrow. The female appears to do most of the work, however, for she is invariably at the far end, with the unchivalrous male behind her. The gribble is a little creature resembling a woodlouse, but only about 3 mm long: it has a semi-cylindrical, segmented body, and seven pairs of short legs, each of which ends in a sharp, curved claw by means of which its owner grips the sides of the burrow. Behind these legs are five more pairs of limbs, each carrying two broad plates, which act as gills, keeping up a continuous movement and pumping fresh

Tooth & Claw

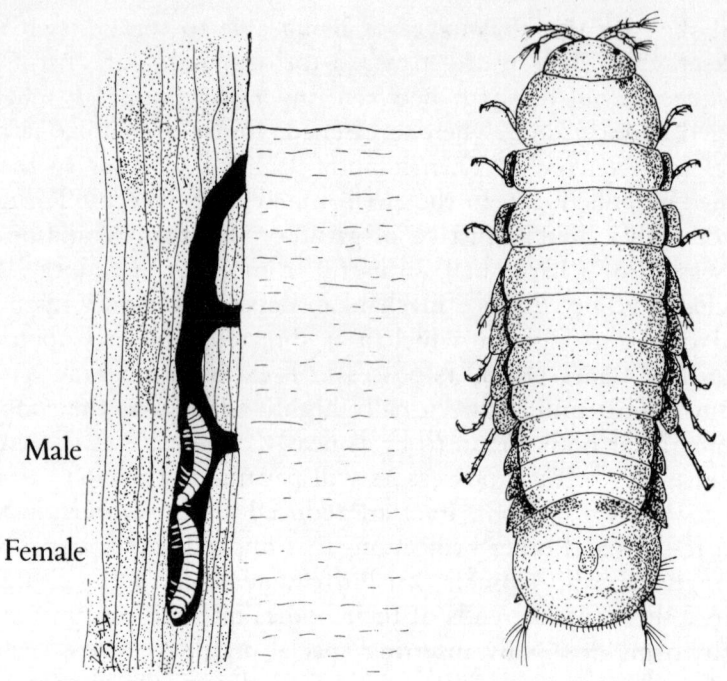

3. The gribble *(Limnoria)*. *Left*, diagram to show the method of burrowing; *right*, much enlarged (actual length 4.5 mm).

water into the burrow. The mandibles are not identical: the right one has a sharp point and a roughened edge which fits into a groove of the left. The latter has a rasp-like surface, and the two act in combination as a file and rasp when boring into wood. The path of the burrow may reach up to 20 mm in length and can be detected from above because its roof is perforated by a regular series of tiny man-holes, or respiratory pits, which help in maintaining the circulation of water within the burrow.

The ravages of gribbles are readily apparent on jetties and pier piles throughout the world, showing how successful they must be in avoiding enemies. As the outer layers of wood are destroyed and fall away, the gribbles are able to destroy progressively more of the structure. The individual burrows of gribbles are the same width throughout, so it seems that the animals must come out when they moult and grow in size, and then start new burrows. The young are incubated in a special brood pouch beneath the body of the female. When they emerge, they begin to

Life in a Hole

hollow out little burrows of their own from the sides of their parents' burrow.

Unlike the gribble, the shipworm *(Teredo)*, a bivalve mollusc, bores by rasping wood with a shell whose front edge bears rows of fine, sharp teeth. The shell is attached to the anterior end of a long naked body, while two delicate siphons at the rear are the only parts of the animal ever to project from the burrow. These are instantly withdrawn if they are touched, and the opening of the burrow – which is no larger than a pinhead – is then closed by means of a pair of club-shaped plates or 'pallets' attached to the posterior of the shipworm, near the bases of the siphons. After a larval shipworm has alighted on a suitable wood surface, it metamorphoses and immediately starts to burrow. However many animals there may be in a piece of wood, their burrows never run into one another although X-ray photography reveals that they may become interlaced in the most intricate manner. Once encased in its burrow, no shipworm can ever leave it, because it grows so much larger than the tiny hole through which it originally entered. Nor does it ever meet another shipworm: eggs and sperm are merely discharged into the sea through one of the siphons during spring and summer, and fertilization takes place in the open waters.

The habit of boring into rock is widespread among invertebrates, including no less than seven superfamilies of Mollusca. Some molluscs, such as the piddock *(Pholas)*, adopt the method of the shipworm while others, including the date-shell *(Lithophaga)*, dissolve chalk and limestone by secreting acid. Other kinds of animals that bore into stone and rock are sponges, worms, and crustaceans. They must all do this solely for defence against predatory crabs, fishes and so on since, unlike woodborers, they cannot obtain any food from their effort.

Many crabs and molluscs simply embed themselves in the sand or mud, but most anchorites make a burrow. Burrows may be straight or curved, simple or complicated in form. The shipworm covers the inside of its tube with a chalky secretion, some of the burrowing polychaete worms secrete linings of various organic substances, while earthworms plaster the sides of their holes with material from their castings. Such linings probably prevent individuals of the same species from inadvertently breaking into the burrows of their neighbours. Some female wood-boring beetles make relatively wide tunnels and place their eggs in niches along the sides. On hatching, each little larva bores for itself a tube at right angles to that of its mother and parallel with those of its brothers and sisters. These larval tubes enlarge from one end to the other as their occupants grow in size, thus forming elegant, and characteristic patterns inside the bark of trees.

Like the shipworm, certain terrestrial anchorites seal their burrows with specially modified portions of their bodies. This trait, which is termed 'phragmosis', is also found notably among certain ants and termites, in which the heads of the soldiers are adapted to fit openings in woody plants by which the insects enter. These ants use their heads, like the thick door of a safe, to close the entrance to the nest and keep out intruders. When a worker needs to go out and forage, she merely strokes the soldier's abdomen with her antennae and the animated door moves back to let her pass. On returning, she taps with her antennae on

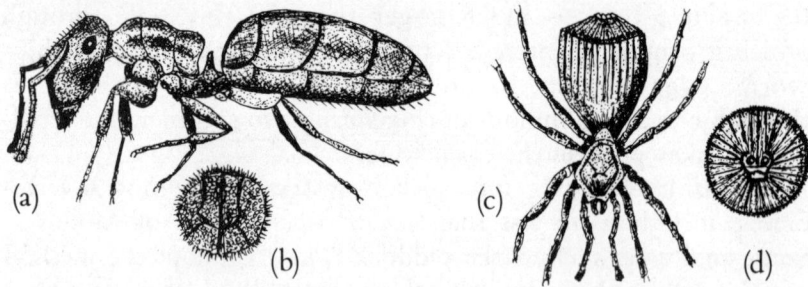

4. Phragmosis, illustrated by an ant *(Colobopsis):* (a) soldier, (b) head of soldier from in front; and by a spider *(Chorizops)* (c), showing (d) truncated end of the abdomen. (After W. M. Wheeler.)

the truncated surface of the janitor's head and is immediately re-admitted! A truncated head or abdomen has been developed, quite independently, in several different types of burrowing animals including worms, insects, arachnids, frogs, snakes and mammals. A particularly dramatic example is afforded by a tropical American trapdoor-spider *(Chorizops)* which uses the tapered posterior of its abdomen to close the underground silken tube in which it dwells.

I have already indicated that the driving force of evolution is the selection of adaptations that achieve their functions with the

Life in a Hole

greatest economy, and this is as true of the evasion of enemies – a form of passive defence – as of any other aspect of animal life. When an animal inhabits an environment in which it does not require any particular portion of its anatomy, we may be sure that that part will soon atrophy, become vestigial, and eventually disappear, like the eyes and colours of animals that live permanently in dark caves. The same is true of cryptozoic animals inhabiting leaf litter, the spaces beneath rocks and the bark of trees. These are mostly inactive, slow-moving creatures that spend long periods of time in a passive state. Their visual organs are poorly developed, an auditory sense has seldom been demonstrated whilst, on the other hand, tactile and taste sense organs are well developed. Efficient respiratory systems are either absent or, when present, lack devices that control water loss by evaporation. These are not required, of course, in a sheltered, moist, environment, but their absence prevents most cryptozoic animals from emerging from their retreats during the daytime.

The family history of the cryptozoa has greater continuity, and reaches back longer in time, than does that of the present inhabitants of more open environments. Many orders of cryptozoic arthropods were already well established when the first fossil-bearing sediments were deposited and only a minority has failed to survive until the present. All this suggests a community of animals which, in former times, was the product of an environment more simple and uniform than the world of today. Evolutionary progress has subsequently died away, leaving the cryptozoa imprisoned in a constant environment where natural selection has reached a dead end.

Of course, these generalizations oversimplify the situation. Living in a hole, a burrow, or any other cryptozoic environment, serves purposes in addition to that of avoiding predatory enemies. Indeed, as mentioned above, many cryptozoic animals are themselves carnivores, preying upon their cryptozoic compatriots. At the same time, however, protection from desiccation is achieved by inhabiting such environments. This is especially important to small desert animals, the majority of which dwell in holes.

Conditions within a hole or burrow tend to be remarkably constant. There is perpetual darkness, high humidity and a temperature which varies only slightly, and with considerable time-lag, from night to day. How then do the anchorites that normally emerge at night become aware that darkness has fallen in the outer world? The simple answer is that their activities are rhythmic: they are controlled internally or 'endogenously', and are manifestations of the workings of the 'biological clock'. Life in a hole lacks external stimulation, so the initiative must come from the anchorite itself!

Earthworms spend the day in underground burrows, but come out at night and crawl around on the surface of the earth, where they mate, or feed on fallen leaves. Even under constant laboratory conditions, they maintain this rhythm of activity for weeks. In addition, worms apparently possess biological clocks with an annual frequency. During the summer, some species remain underground in a state of suspended animation. Although the garden earthworm *(Lumbricus)* does not aestivate in this way, its reproductive organs become mature in spring and it mates only during the summer – thereby demonstrating either that it possesses a seasonal clock or else that it responds to seasonal changes in the environment. This rhythm is paralleled by another which ensures that earthworms move especially quickly on summer evenings when they crawl outside their burrows and may need to escape rapidly if discovered by an enemy.

Diurnal rhythms of activity have been studied quite extensively in woodlice. These remarkable terrestrial crustaceans are cryptozoic and spend the daytime under stones, bark, fallen leaves and in other damp, dark places. They tend to dry up very readily, however, and only come into the open after nightfall, when the temperature drops and the relative humidity of the air increases. Not only do their biological clocks regulate this activity, but they trigger a decrease in the intensity of the response to humidity: this enables woodlice to walk at night in quite dry places where they are never to be found during the day. At the same time, an increased avoidance of light after being in darkness for a while, ensures that they get under cover promptly when day breaks – if they have not already retired before then – and thus they avoid the attentions of early birds and other day-active predators. Nocturnal habits, controlled by the biological clock, are important to woodlice for survival in a hostile environment populated by predatory enemies.

To the student of nature, the lives of animals that inhabit burrows and holes raise quite fascinating physiological and ecological problems, the answers to which are still by no means fully understood. We cannot yet begin to comprehend the mechanism of the biological clock, nor how it keeps time when the weather changes – most biological processes speed up when the temperature rises. It is not known how the reactions of woodlice to light and moisture are coupled with their biological clocks, how anchorites respond to daily or seasonal changes in the outside world, nor how the sexes find one another in the mating season. On the other hand, we can see how animals both gain and lose through adopting a life of seclusion. Adults may enjoy comparative immunity from predatory enemies but, in achieving this, they forfeit a considerable degree of mobility. Dispersal becomes the responsibility of the larvae or young which are small and vulnerable and, therefore, must be produced in very large numbers if any of them are to have a reasonable chance of survival. An adult animal has to pay for the security of a hole-dwelling existence by producing more offspring than would be necessary if it had not abdicated so much of its parental responsibility.

2 Camouflage

An anchoritic existence is not always the answer to the problems of survival in a dangerous world. Nor, in human warfare is conflict resolved by trenches and bomb-proof shelters alone – indeed, not until the invention of machine guns and high explosive shells was it even advisable for soldiers to dig holes and trenches! Other defensive strategies are also necessary. During the Great War, the development of photography and aerial reconnaissance made essential the concealment of ammunition dumps, gun batteries, troop concentrations, ships, factories and other objectives of military and economic importance. These were camouflaged in various ways: sometimes their outlines were obscured with splashes of various colours, or screened in order to eliminate shadows. (Sometimes also, potential targets were disguised so that they could not be recognized by the enemy for what they actually were.) Over the years, with the growth of increasingly sophisticated technology, camouflage has become correspondingly more sophisticated and today includes such ploys as the jamming of enemy radar. Even this may have its counterpart in the animal world.

Camouflage (like disguise, which we will be discussing in the next chapter) is an important defensive mechanism of animals and employs principles which, not surprisingly, are often similar to those used in human warfare. The basic type of camouflage is seen in simple environments, such as snowland, tropical rainforest and desert, whose inhabitants are often coloured uniformly so that they match their backgrounds. In such cases, camouflage may be used in offence as well as in defence, so that predators like snowy owls and desert foxes, as well as their prey – lemmings or jerboas – have 'cryptic' or concealing coloration.

At other times, as with human warfare, camouflage in the

animal kingdom can take a number of different, highly sophisticated, forms. Any solid object is normally presented to the eye merely as a patch of colour occupying a particular area in the visual field, and the ways in which it can be distinguished are fourfold. First, the colour of the object may differ from that of its immediate surroundings – a black beetle is conspicuous on yellow sand, but a yellow insect is not. Secondly, the object may be thrown into relief by the effects of light and shade which enable the eye to detect surface curves and texture. Thirdly, the form of the object may be recognized by its outline or contour and, fourthly, the object may cast a distinctive shadow. From this, it follows that, for camouflage to be effective, it must consist of one or more of the following: colour resemblance between the object and the background against which it is seen; 'obliterative counter-shading' to eliminate the appearance of roundness due to light and shade – many cryptic animals are pale on the ventral side and darker in colour above; elimination of shadows by flattening the body and orientation towards the source of light; and 'disruptive coloration', a superimposed pattern of contrasted colours and tones which breaks up the surface outline, as do the stripes of a zebra or tiger.

The ability of animals to achieve concealment, even though they are fully exposed, and despite seasonal changes in their environment, is sometimes quite remarkable. A ptarmigan, sitting motionless upon its nest among the lichen-covered rocks on a barren mountain-side, may be almost invisible unless it is inadvertently disturbed and takes flight. The ptarmigan is a grouse which inhabits high, mountainous slopes. At all seasons of the year, it has white wings and a white belly but, in the northern part of its range, it becomes pure white in winter – with the exception of the black tail which, however, is largely hidden by white tail-coverts when the bird is at rest. Most birds of temperate regions begin to moult their summer dress as soon as the young are fledged: in the Arctic, the process has to be accelerated because the summers are so short. In Iceland and the far north of Europe and Asia, the ptarmigan moults no less than three times between June and September – from winter to summer plumage, from summer to autumn plumage, and back to winter white! Several other northern animals also change to white at the approach of winter. These include the willow-grouse, mountain

Tooth & Claw

5. Ptarmigan in summer and winter dress.

hare, Arctic fox and stoat. The latter changes not only its pelage but also its name: in the winter phase it is known as 'ermine'! Again, the degree of seasonal change often varies with latitude and climate. For instance, the stoat, which always turns white in the north of Scotland and on Ben Nevis, does so less completely and less frequently in other parts of the British Isles. In the extreme north of its range the Arctic fox nearly always discards its brown summer dress for a pure white winter coat while in Iceland, where the winters are less severe, a change to white is exceptional. In North America, the prairie-hare and snowshoe-rabbit turn white in winter; the polar-hare is white all the year round; while the wood-hare, which ranges further south, never loses its summer coloration.

In tropical forest, green shades invariably predominate, rendering many forest animals invisible against a background of leafy foliage. Frogs, and reptiles such as geckos and tree-snakes are usually green in colour, while the same is true of forest insects such as grasshoppers, caterpillars, and plant-sucking bugs. Among forest birds also, green is a very common colour. This predominant green coloration – and, of course, all animal coloration – can have a number of different causes. In caterpillars it is usually green food in the alimentary tract showing through a transparent body wall. In other insects, green may be due to modified chlorophyll or be synthesized independently of the nature of the food – as are the green pigments of marine and fresh-water crustaceans. The green colours of mature insects are often caused by the refraction of light on the surface of the insect's body – a metallic sheen is due to the fact that a large proportion of the light is reflected.

Most vertebrates owe their colours to specific chemicals, but the colours of feathers can be either chemical, physical or a combination of both. In most birds, green is produced by a combination of melanin – a black pigment – and a yellow pigment. In the turacos or plantain-eaters of tropical Africa, it is due to turacoverdin, a peculiar pigment the nature of which is not yet known. The amount of this green pigment is directly related to the luxuriance of the vegetation in which the birds live, being most developed in the numerous species of evergreen forest, and least or not at all in deciduous environments, especially thorn scrub.

Desert animals, whilst in general they tend to differ from their relatives elsewhere in the possession of a buff or sandy coloration, are not simply coloured fawn, brown, cream or grey indiscriminately. There is often a very close similarity between an animal and the particular kind of soil in the desert where it lives. Moreover, certain reptiles, such as chameleons and lizards, are able to change colour at will to match their changing backgrounds. Many African butterflies show seasonal colour forms, which appear at different times of the year. During the dry season, predation on butterflies is most severe and natural selection exacts its heaviest toll, so the brown, dry-season colour is the one best adapted for concealment. It is well known that ground-nesting birds tend to have plumage colours and markings that conceal them from view when seen against the soil on which they live. This is especially true of desert species, such as crested larks, whose native terrain provides only sparse cover. It also occurs in mammals, including nocturnal species such as gerbils and jerboas – for, in desert regions, the atmosphere is usually exceptionally clear, and distant objects can be seen quite clearly when a flood of silver moonlight bathes the scene.

The ability to change colour is not confined to chameleons and lizards; crustaceans, cuttlefish, octopuses and fishes are also able to adapt to different backgrounds. Colour change often takes place extremely rapidly: cuttlefish and octopuses can change in less than a second. When swimming, the common Atlantic cuttlefish usually adopts a zebra-striped pattern which breaks up the outline of its body but, when resting on the bottom of the sea, it immediately adopts a sandy or mottled colour which matches the sand and pebbles beneath it. Colour changes in prawns, flatfish and chameleons may be equally striking, but take very much longer to achieve. The speed of change is related to the methods evolved by different animals – in the cuttlefish and other cephalopods colour change is under direct nervous control which brings about rapid muscular contraction of pigment granules in special cells, known as 'chromatophores'. In other animals, however, the chromatophores are branched cells in which pigment is slowly dispersed or concentrated in response to hormonal action.

Many of the animals that inhabit the transparent surface waters of the sea elude the attentions of predatory enemies by themselves becoming transparent. This feature is shared, in the larval or adult state, by widely different pelagic organisms – including unicellular protozoans, jellyfishes, ctenophores or sea-gooseberries, bristle-worms, molluscs, crustaceans, sea-squirts and fishes. The transparency of pelagic organisms has been remarked upon by many naturalists, but little is yet known either about the physical means through which it is achieved, the visibility of the organisms to their predators, or the part which transparency may play in their lives. The degree of transparency varies greatly from one species to another – many jellyfishes, although partially transparent, can readily be discerned by the human eye. In contrast, ctenophores are completely invisible in the water – although they may easily be caught by touch. In their case, crypsis may be useful in offence as well as defence, since sea-gooseberries are themselves predatory organisms.

Despite its advantages, transparency seems to be something of a mixed blessing. Since camouflage which results from matching the colour and pattern of the background is effective only when the animals concerned are motionless, this type of concealment would be completely useless to marine animals which must either swim or float at the mercy of the waves. The advantage of a transparent body over one which is camouflaged, like that of most fishes, by reflective scales, is that it affords concealment against any background and, at the same time, permits movement and locomotion. The disadvantage, on the other hand, would seem to be that transparency has to be achieved at the expense of having a bulky, gelatinous structure, so that the animal can only drift or move very slowly. The last word has clearly not been said about the usefulness of transparency, however, since, for example, one of the siphonophores responds to touch by becoming milky-white within a second or two, and does not regain its transparency until nearly half an hour has elapsed: the significance of this is not known.

Animals which are not transparent can only conceal their natural shadows by the device known as 'obliterative counter-shading'. A counter-shaded animal's back is usually of a deeper hue than the flanks, which grade to a pale under-surface so that shadows are not apparent. In many desert mammals especially, the underside is very pale or even quite white. Moreover, the pale, ventral area is often extended over the flanks in desert forms to a greater extent than it is in related species from other habitats where there is more cover and shadows are less conspicuous. This is found in spiders, centipedes, woodlice, insects, lizards, snakes and birds. Counter-shading is also characteristic of fish and other pelagic marine animals that do not achieve inconspicuousness through transparency.

Silvery fishes are better camouflaged than are normally counter-shaded fish. Sharks and dogfish, for instance, have typical counter-shading. When viewed from above, or from any angle above the horizontal, the dark dorsal surface reflects about the same amount of light as is reflected up from the depths of the sea so that the fish is camouflaged; but, from any point below, the outline of a shark of dogfish appears as a dark silhouette. A silvery fish, such as a herring or mackerel, is not only camouflaged when viewed from above, but reflects light when viewed from the side. It can only be seen easily by a predator looking directly from below, and even then the silhouette is minimized by the lateral flattening of the body and the tapered ventral keel.

Some deep-water fishes, crustaceans and squids, camouflage their silhouettes, when viewed from below, by means of batteries of

Tooth & Claw

uniformly spaced luminescent organs or 'photophores', which direct light ventrally. It has been shown experimentally that the amount of light emitted is almost identical with that found at the depths at which these animals normally live, so that they are concealed very effectively from below. At the same time, their backs are black and their sides are reflective so they are equally well camouflaged to predators from all directions. One marine shrimp *(Sergestes)*, which lives at depths between 25 and 400 metres, is transparent apart from its digestive organs, eyes and photophores. The reason why the alimentary canal needs to be opaque, and therefore luminescent, is that the prey themselves are bioluminescent, and a stomach full of gastronomic delights would immediately betray the shrimp that had gathered them.

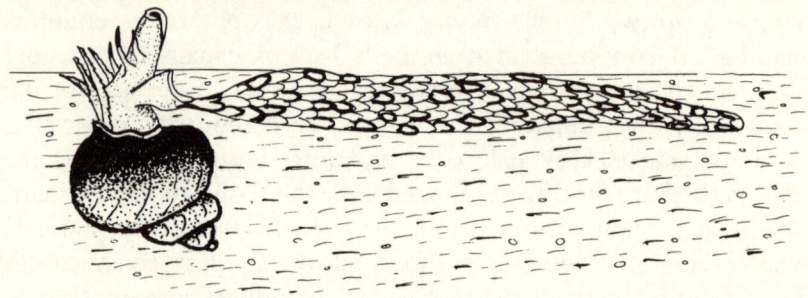

6. The purple sea-snail *(Ianthina)*, a pelagic winkle, hanging beneath its transparent float of hardened slime enclosing air bubbles.

Animals which normally rest upside down usually show reversed counter-shading, with the dorsal surface pale and the ventral, which is now the upper, surface pigmented. Well-known examples include puss-moth and other caterpillars, various spiders, some species of catfish – of which one example is a common inhabitant of the Nile – tree-sloths, and the pelagic winkle *Ianthina*. This last, a tropical mollusc common in Indo-Pacific waters, hangs upside down from a bubble float of hardened slime. Its body and the ventral surface of the shell are deep purple in colour, the upper whorls are pale.

Another source of conspicuous shadows is lateral illumination, but animals often eliminate shadows caused in this way by pressing their bodies against the ground or the substrate on which they are resting. Such behaviour is characteristic of creatures that live on tree trunks. For instance, the flying gecko *(Ptychozoon)* of Malaya is not only cryptic, with disruptive coloured markings on the body and tail, but its shadow is concealed when at rest by the flattened shape and tapering flanges on the tail which eliminate shadows. Arboreal insects, such as mantids, cicadas, grasshoppers and caterpillars are also often flattened in shape so that lateral shadow is eliminated.

Ghost-crabs, which scurry so silently across tropical beaches, owe their relative invisibility not so much to their sandy colour as to the fact that they shelter instinctively in natural hollows where their shadows are concealed, or they rapidly excavate shallow holes that serve the same purpose. Similar behaviour is characteristic of the young of birds such as ostriches, plovers and curlews, which squat motionless with necks outstretched so that only the cryptic upper surface is visible, while shadow is almost obliterated. Likewise, many of the smaller African antelopes crouch down or lie with their necks stretched on the ground when they are alarmed.

It may seem initially surprising that the black-and-white stripes of the zebra should be an efficient style of camouflage, but this is an excellent example of the use of disruptive coloration. Herds of zebra are very easily seen by day in the open country or bushy scrub of their natural home but, since they are fleet of foot, with acute senses of sight, hearing and smell, they are among the most wary and difficult of wild game to approach during the hours of daylight. At dusk and dawn, however, when they are especially liable to attack by lions and leopards, their black-and-white stripes blend into an elusive light grey which matches the dim background, and the animals become almost invisible. Their outlines melt away, for the dark stripes cut right across the outer contours at every part of the body while, on the underparts which would naturally tend to be shaded and rather darker in colour, the dark stripes are narrower than elsewhere. So the bulky body, instead of looking solid by reason of natural shade, appears flat in consequence of its counter-shading.

The striking dazzle pattern of the zebra's camouflage, which breaks up its outline and renders the animal almost invisible at the time when it is most vulnerable, is but one instance of many. Of course, camouflage cannot be completely effective at all hours of the day or night, but nature has ordained that it may be so at the times of maximum predation when natural selection is most severe. Indeed, a particular colour pattern may well have been selected, even if it is only protective at one important period in the life of its possessor, such as during the breeding season.

Disruptive colour patterns are found throughout the animal kingdom. They are to be seen, for instance, in the pale, dorsal stripes of different grasshoppers, mantids and caterpillars, in the irregular blotches of dark colour that characterize the skins of amphibians and reptiles, in the stripes of weed-dwelling fishes, and in the camouflaged fur of forest mammals. A fine example is provided by the buff and black eggs of the lapwing or, even better, by those of the stone curlew, whose outlines are so broken up that they are almost invisible. These are in striking contrast to the white, unpigmented eggs of birds whose nests are hidden in holes. Moths that normally rest on tree trunks with the head upwards have disruptive marks running parallel to the body axis, while species that normally rest sideways have disruptive marks running transversely. Whether or not the outline is broken up optically, it is usual for the disruptive pattern to extend as an irregular patchwork of cryptic tones over all parts of the body that are normally exposed to view.

The eye is one of the most difficult parts of the body to conceal on account of its regular, sharply defined and rounded shape. Among certain nocturnal geckos and snakes, whose pupils are contracted in daylight to mere slits, the iris is of a tint which matches the rest of the head. Most vertebrates, however, have large, round, black pupils, which have to be camouflaged by extended irregular patches of dark pigmentation running across the head, as seen, for example, in various species of fishes, frogs, the ringed and Kentish plovers, and the oryx. While this is the most widely accepted explanation of the function of eye stripes, it has also been pointed out that the birds which possess them are frequently those that feed on fast-moving prey. Since the eye-stripe normally passes from the eye to the beak, it could act as a sighting line to increase the chances of successfully catching prey. In predatory fishes, and other animals, however, where the stripes do not pass from the eye to the mouth, their function can only be one of camouflage.

The development of projections and processes that break up its outline are another means by which the body contour of an animal can be concealed. This is characteristic of many of the most perfectly camouflaged species. For example, the outer wings of the comma-butterfly and of many moths are highly irregular and render the insects very inconspicuous when they are at rest; and deceptive modifications of a similar kind occur in various grasshoppers, leaf-insects, and plant-bugs where they intergrade with modifications representing definite features of the environment – leaves, twigs, thorns and so on. Such camouflage, therefore, incorporates the principles both of disruptive colour pattern and of disguise, the subject of the following chapter.

Concealment or 'crypsis' is probably one of the most important ways by which enemy attack can be avoided, but it is not restricted to visual camouflage and, for that matter, disguise. Many carnivorous animals hunt by scent or sound, and these features also need to be camouflaged both by predators and by their prey. The writings of big-game hunters contain frequent references to the remarkable way in which the larger cats, such as leopards and tigers, stalk their prey upwind. Antelope and other game have a keen sense of smell, so that the enemy must mask his scent if he is to approach within striking distance. Lions do the same when hunting singly but, when together in a pride, they adopt more sophisticated tactics by which most of them approach upwind while one or two stalk from the flank so that their scent drives the prey into an ambush. At the same time, it must profit prey species not to produce a strong smell that would betray them to their enemies.

Whilst many animals protect themselves from attack by possessing a repugnant smell, very little is known of the ability to avoid enemies either by being relatively odourless, or by adopting a false scent – although this almost certainly takes place. In recent years it has become increasingly apparent that the use of 'pheromones', or scents for intra-specific communication, is widespread and often extremely sophisticated, so it would be most surprising if scent could not sometimes be masked, or even used to mislead potential enemies. Colour crypsis has often been studied in animals, because it is comparatively easy to observe the responses of predators that hunt by sight to prey of different colours. Presumably, however, animals can also be cryptic with respect to predators hunting by the sense of smell, hearing or any other of the senses.

Just as predators that hunt by sight search for a particular visual image, so the predators which hunt by smell may seek out a particular odour and ignore potential prey that are giving out a different scent. One of the few experimental investigations on this topic was carried out using laboratory mice as 'predators', and 'prey' consisting of artificial baits coloured identically but scented with either peppermint essence or vanilla. After three days of training, during which groups of mice were exposed to 20 baits of one variety for periods of ten minutes, animals conditioned to peppermint took significantly more bait with this odour, while mice conditioned to the smell of vanilla took significantly more of the vanilla-scented bait. The mice were therefore taking more 'prey' with a familiar, than with a novel smell. If prey whose predators hunt by scent were common, the best defence against them would be the evolution of a variety of different scents within the prey species. In a similar way, colour polymorphism – the evolution of many different colours within the same species – is frequently evolved among relatively defenceless animals whose enemies hunt by sight.

The avoidance of detection through sound is equally important to the potential prey. Silence is relatively unimportant in open country, of course, because enemies can be seen from afar but, in forest or tall grass, no animal can afford to draw attention to itself by making unnecessary noise. It seems probable that the high frequency sounds used for social communication by rodents may well have been evolved because predators such as owls, which possess extremely sensitive hearing at lower frequencies, are unable to detect them. Ultra-sound attenuates rapidly: it can only be used between animals in relatively close contact with one another, and does not carry far enough to be a positive danger. Moreover, even if a predator could detect ultra-sounds, it would be extremely difficult to locate their source.

Silence is a two-edged weapon and can be used for offence as well as in defence. Most owls are remarkable for their slow, silent flight: their wings are extremely large for their weight and the breast muscles smaller than in any other group of birds, giving great economy in energy. Silence is also enhanced by the arrangement of the feathers. This not only enables owls to approach their prey unnoticed and surprise it, but makes it possible for them to detect any noise, however faint. The most outstanding skill in the localization of prey by sound is seen in these birds. A barn-owl, flying in a completely dark room, can catch a mouse running on the floor. Owls differ from bats in not using echo-location, nor do they perceive ultra-sound, but their ability to localize audible sound is very high. Some owls have developed an asymmetrical arrangement of the ear cavities and asymmetrical external ears to assist in this.

Like all defensive mechanisms, camouflage cannot be perfect. It is impossible to elude every enemy, but the predator most commonly encountered can often be deceived. When different predators make use of different senses to detect the same prey, the latter may have to be camouflaged in more ways than one. Natural selection results in a compromise between conflicting requirements – the need to be invisible and silent to avoid detection and, at the same time, to move about, feed and communicate with other members of the species.

3 Disguise

During the approach march preceding the Second Libyan Campaign which led to the relief of Tobruk, the tanks of the British 7th Armoured Division were disguised with canvas screens so that they looked like lorries. From the ground the deception was reasonably effective, but no doubt the track marks in the sand would have immediately disclosed the deception to aerial reconnaissance. On the first day of the battle, 18 November 1941, the 4th County of London Yeomanry (Sharpshooters), supported by 2nd Royal Gloucestershire Hussars, attacked Bir El Gubi, just inside 'the wire' – an immense barricade of barbed wire erected by the Italian army. In retrospect, the wisdom of this action may be doubted but, at the time, it was not realized that the enemy position was heavily fortified, and that the stationary column of what appeared to be transport vehicles on the Italian left flank was actually composed of dummy trucks, with anti-tank guns dug in behind them! In the resulting mêlée, about 45 British tanks were knocked out.

Certain animals, similarly, use disguise to tempt their prey within reach. The terrible jaws of the angler-fish are concealed by a curtain of innocent-looking weed-like lobes. At the same time, the fish also carries on its head a line and 'lure' which attract investigation by the small inshore fishes upon which the angler feeds. The apparatus is situated well forward on the snout and consists of an elongated, flexible ray of the first dorsal fin bearing, at its extremity, a flattened, flag-like appendage which serves as a 'bait'. In the deep-sea anglers, the lure takes the form of a bulb which can be made luminous at will and acts as a lamp to attract other fishes to destruction. The ultimate is found in certain angler-fish *(Antennarius)* whose lure not only mimics a small fish, and provides a highly attractive visual cue, but also creates a low-frequency pressure stimulus. The thin, membraneous quality of the bait allows it to ripple while being pulled through the water, thereby simulating the lateral undulations of a swimming fish, and generating pressure waves which attract the predators that form the food of the angler.

Among reptiles, a comparable example of aggressive disguise occurs in the matamata *(Chelys)* of Brazil, a cryptic, bottom-dwelling turtle, which bears beneath its mouth a series of red filaments that look remarkably like small worms. When a hungry fish or frog is attracted to one of these, it is immediately sucked into the cavernous mouth of the matamata whose jaws are weak and are not used for seizing prey. The alligator-snapper *(Macroclemys)*, a powerful, mud-coloured turtle, which inhabits the rivers of North America, does something similar. Its dull yellow head and shell match exactly the coffee-coloured waters of the lower Mississippi, where it lies motionless on the bottom. It lures its prey to destruction with a curious attachment on the inside of its lower jaw, close to the tongue, which looks like a large white grub crawling about in the mud.

One of the most extraordinary examples of the use of disguise to deceive prey is afforded by the African water-mongoose which relishes a diet of birds' flesh. To capture its favourite prey, the mongoose is alleged to resort to a most peculiar trick. It conceals itself in grass or other vegetation, raises its rear end, and distends the anal orifice to such an extent that it resembles a flower or ripe fruit. Unsuspecting birds are thereby lured to peck at the mongoose which adroitly turns about and immediately seizes them. This story is prevalent among so many African tribes that it seems highly probable that it may be based on fact. Stranger things can happen, and the wiles of mongooses in encompassing the downfall of their vicitims are well known.

Disguise, however, is used by animals much more often in defence than it is for deceiving prey. Many moths, whose outlines are broken up by projections and fringes of the wings and body, tend to resemble the leaves of plants, sticks, and other inedible objects of no interest to potential carnivorous enemies. Other insects that resemble leaves or stems include leaf- and stick-insects, grasshoppers and butterflies. Indian stick-insects, which resemble the brown grass stalks among which they live, sometimes heighten the illusion by allowing themselves to be

Tooth & Claw

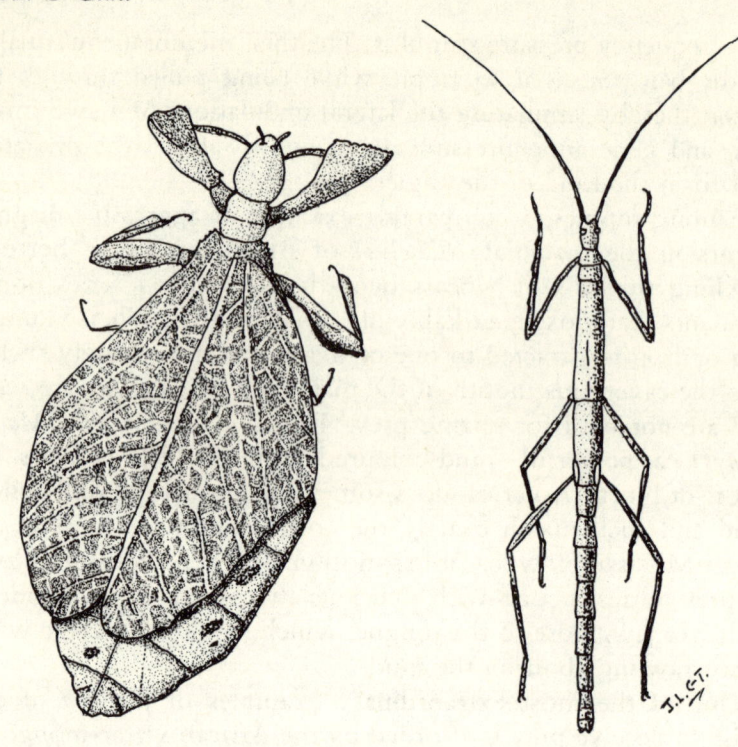

7. A leaf-insect and a stick-insect, both members of the same insect family (Phasmidae).

blown about by the wind in exactly the manner that the grass itself is blown.

Protective resemblance, the mimicry by an animal of an inanimate object, or of an inedible part of a plant, is a widespread defensive device among insects. Other familiar examples include the mantids and geometrid caterpillars that mimic twigs and sticks, while many butterflies rest with their wings closed so that they look like dead leaves. Tenebrionid beetles of the genus *Cossyphus* found in tropical Africa, Asia and the Mediterranean region of Europe, like members of the Australian genus *Helaeus*, mimic winged seeds. The wing cases of the beetles are greatly extended so that the insects have a flattened oval shape, while the body is only slightly thickened on its ventral side. Until the beetle moves a limb or antenna, it can scarcely be recognized for what it is. Several species of tree-hoppers bear a striking resemblance to stipules or thorns. The likeness is achieved by the strange

Disguise

shape of the insect's thorax which is greatly enlarged, sometimes covering the whole of the body. In some species, the thorax assumes a quite fantastic shape. Resemblance to a thorn is also achieved by caterpillars which construct somewhat curved, conical cases, made of brown vegetable hairs. They live in them, clinging through an opening in the base of the cone to the plant on which they feed.

The chief difficulty in resembling a leaf, in so far as an animal is concerned, lies in achieving the appearance of being very thin. Most leaves are necessarily thin so that they present a large photosynthetic surface to the rays of the sun but, among animals,

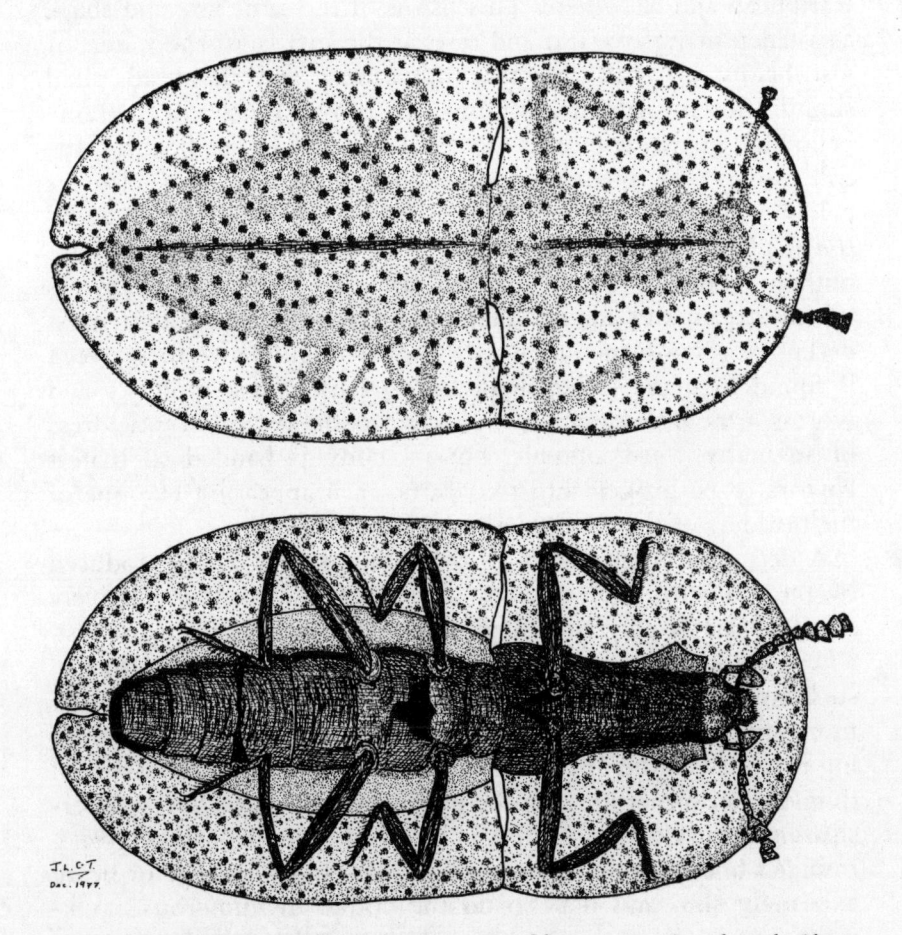

8. *Cossyphus* from East Africa, showing resemblance to a winged seed. *Above*, dorsal view; *below*, ventral view.

extreme thinness is rare and can be attained only at the cost of profound modification of the body and its internal organs. The body is severely compressed in the case of leaf insects, chameleons and some toads. It is also much flattened in certain fishes, which look like fronds of waterweed. One of the Amazon fishes bears a remarkable resemblance to a dead leaf. Only at home in water that is quiet and still, it drifts about, hanging head downwards, or rests on the bottom lying motionless on its side, where it is lost to view among the water-logged foliage and weeds.

Another fish that mimics leaves lurks among mangroves in the Caribbean and elsewhere. This fish is of the same size and shape as a fallen mangrove leaf and rests at the surface of the water, or just below it, with one side uppermost and the head tilted slightly downwards. It has numerous brown spots and the same yellowish-brown colour as a dead leaf: if real leaves float by, the fish moves slowly towards them and, when necessary, changes colour slightly to perfect the match. The so-called 'sea-dragon' *(Phyllopteryx)* of Australian waters is really a sea-horse whose outline is distorted and broken up into numerous filaments which stream out in the surrounding water and create a most deceptive resemblance to weed. Special resemblance to seaweed is found in many other small fishes, crustaceans, molluscs and worms. One little prawn not only wears the typical cryptic dress of so many weed animals, but its body is banded so that it appears to be broken into two parts, each appearing like one of the bladders of the *Sargassum* weed in which it lives.

A deceptive resemblance to the thinness of a leaf is produced by the wings of some insects whose bodies are either relatively small, or are largely concealed when their possessors are at rest. Thus protected by a more or less detailed resemblance to foliage, such insects usually settle with the four wings extended flat. The more specialized among them even have a false mid-rib extending right across the upper surface of the wings. An impression of thinness can sometimes also be produced by obliterative countershading. Many tropical forest lizards and chameleons have solid, rounded bodies which nevertheless give the appearance of being extremely slim and flat. So do the bodies of numerous hawk-moth larvae, in particular that of the eyed-hawk, in which a wonderful degree of optical flattening is achieved by a combination of inverted counter-shading, an inverted attitude, and a superimposed design representing the effect of light and shade on lateral leaf veins.

A different method of achieving apparent flatness depends upon the development of folds of skin – as is sometimes found in tropical toads – or upon ridges of cuticle which serve to sharpen the edges of the bodies of forest grasshoppers and other insects, thus making them appear very thin and leaf-like. Finally, some of the animals that mimic leaves overcome the problem of acquiring a slim appearance by resembling a curled or crumpled leaf. This is a favourite device of tropical butterfly larvae and pupae.

The achievement of a leaf-like appearance, then, requires considerable modification of shape: it is naturally very much simpler to acquire cryptic resemblance to lichen, bark, lianas, twigs, thorns, stones or excrement. Among bugs, moths, long-horn beetles, weevils, spiders, harvest-spiders and other invertebrates, cryptic resemblance to bark and lichen is extremely common, but this type of disguise is by no means restricted to the invertebrates. Remarkable camouflage of the same kind is to be found among arboreal vertebrates, especially tree-frogs, lizards and birds. For example, the little pine-wood tree-frog of North America bears a brown, bark-like, appearance similar to that of many moths; and tree-dwelling geckos throughout the world frequently adopt a bark-like appearance. Whenever lichen forms a characteristic feature of the environment, a diverse array of different animals will be found to have evolved in association with it. Many tree-snakes are extremely attenuated and vine-like in appearance – indeed, only those who have seen them in their native haunts in tropical forests can appreciate the marvellous way in which the appearance and habits of these snakes combine to create a deceptive resemblance to green tendrils or dead vines, twined loosely among the foliage.

In most of the instances of disguise I have mentioned, the camouflage would be completely useless unless the animals normally rested in the correct environment and this, of course, they naturally do. Furthermore, their concealment would be lost if they were to move about unnecessarily. It is not surprising, therefore, to find that adaptive coloration and disguise have been evolved alongside appropriate patterns of behaviour. For instance, most people will have noticed how a stick-insect, if breathed upon, will sway on its legs like a twig waving gently in the breeze. When the vegetation is moving around it, a stationary object would stand out immediately. To take another example, the eggs of the ringed plover resemble so closely the stones of the pebbly beaches on which they are laid that they are usually extremely hard to discover. They are protected by their disguise from marauding gulls and other animals that would undoubtedly devour them with alacrity if they could be detected. Eggs do not move, but when the babies hatch they, too, are disguised to resemble stones. While they are moving about, they are immediately obvious so, if danger threatens, they respond by 'freezing', and remain completely immobile for long periods. Baby ostriches that are unable to keep up with their mothers, lie flat on the ground with necks outstretched, looking like heaps of pebbles and trusting in their camouflage and absence of shadow to evade detection.

In the Western Desert of Egypt, there are mantids which have no functional wings. These run very fast over the gravelly surface of the desert, whose colour they match exactly. When they stop suddenly, they look just like one of the numerous small stones with which the desert is littered. Many Saharan grasshoppers, too, look exactly like stones when they are motionless. The habit of stopping abruptly makes it almost impossible to determine the exact spot in which they have come to rest.

Many unrelated insects throughout the world disguise themselves as excrement, but the deceptive appearance may be brought about in entirely different ways. Since bird and lizard droppings are often deposited conspicuously on leaves and are avoided by other animals, it is only to be expected that a number of insects should have developed a close resemblance to them. Caterpillars of several species of swallowtail butterfly mimic bird droppings and, while they are small, rest conspicuously on the upper surfaces of flat leaves. When they grow too large to look like droppings, however, they become green and cryptic, and remain on the stems and undersides of the leaves. Some geometrid moths habitually rest with their wings extended and closely applied to the surfaces of the leaves on which they sit. In this position they forcibly suggest the appearance of birds' excrement which has fallen from a great height and become flattened into a wide patch. Small caterpillars sometimes mass together, the whole group simulating the droppings of a bird, and the twisted masses of egg sacs of an Indian scale insect achieve the same effect. Certain tropical beetles, likewise, mimic the excrement of birds, lizards, or even of caterpillars.

A particularly striking example of this type of disguise is provided by a Javanese crab-spider which constructs on the surface of some prominent dark green leaf an irregularly shaped film of web. On this, the spider lies on its back, with its black legs crossed over and closely pressed to its body, thus exposing the white ventral surface of its abdomen. The white and black together look like the central, dark portions of the excreta, while the thin web represents the marginal watery portion becoming dry, even to some of it trickling off, and arrested in a thickened extremity such as an evaporated faecal drop would leave. Although this disguise is frequently cited as an example of alluring coloration – many butterflies regularly feed on birds' droppings – I think it is more probably an anti-predator device. The sharp eyes of a bird are far better able to distinguish between excrement and a spider than are the compound eyes of a butterfly. Furthermore, since insect vision extends further into the ultra-violet than does that of vertebrates, it is improbable that the resemblance to excrement would be so perfect when viewed by an insect rather than by a bird. Similarly, the crab-spiders which change colour to match that of the flowers in which they are hiding, are quite conspicuous when photographed by ultra-violet light. It must therefore be assumed that their camouflage is effective against vertebrate enemies and does not function to hide them from their insect prey.

This may be a convenient point at which to discuss the special protective resemblances of insect larvae, pupae and cocoons. An insect in the pupal or chrysalis stage has, in large measure, broken off relations with the external world. Pupae do not feed, and move little or not at all. Some primitive insect pupae have large mandibles which help the adult to escape from the cocoon, and these can be used in defence but, in general, insect pupae and, to a lesser extent larvae, are relatively vulnerable, and their survival depends on a variety of protective devices. There are two classes of enemies that prey upon insect pupae when opportunity arises, the smaller of these being arthropods, the larger mainly vertebrates: protection against each of which is usually afforded by quite different structural characters. For instance, ladybird beetle pupae have organs called 'gin-traps' – furrows in their integument, separated by ridges and spines which can inflict a sharp nip on attacking mites and small insect enemies: these will be discussed in more detail later. At the same time, however, ladybird pupae are protected from the attacks of vertebrates by distasteful fluids associated with conspicuous, warning coloration.

Protective resemblance, of course, is concerned with misleading large vertebrate enemies. One West African moth caterpillar spins a very thin-walled, golden-yellow cocoon on the upper surface of a green leaf. The cocoon has two compartments, an upper and a lower one, separated by a thin sheet of silk. The pupa is green, like the leaf, and is tucked into the lower compartment. It is quite easy to see into the upper part of the cocoon, which appears to be empty because the green colour of the pupa beneath prevents it from being recognized.

Caterpillars of several unrelated families of moths attach to the surfaces of the pupal cocoons structures that resemble the cocoons of parasites. These structures, known as 'false cocoons', usually look like the cocoons of braconid wasps. It is presumed that birds learn that genuinely parasitized cocoons are almost inedible and so do not bother to open either them or others disguised to look as though they, too, have been parasitized. The larvae of some of the crane-flies of tropical America secrete quantities of a hygroscopic protein from enlarged glands associated with their mouths. This protein absorbs water vapour, swells up around the larva and makes it look like a rain drop about to fall from the end of a leaf. Small tortoise-beetles, too, sometimes look like drops of water. In temperate regions, the nymphs of frog-hoppers secrete themselves in a frothy substance commonly termed 'cuckoo-spit'. By this means they are protected from some predaceous insects and other arthropods. Nevertheless, they are not infrequently seized from their spume by hunting wasps, so the protection afforded can be only relative. No doubt it is an effective disguise as far as vertebrate enemies such as birds are concerned.

Many caterpillars, especially those of hawk-moths, are well known for their resemblance to snakes. The pupa of a Burmese moth *(Tonica)* also bears a very close likeness to the head of a

Disguise

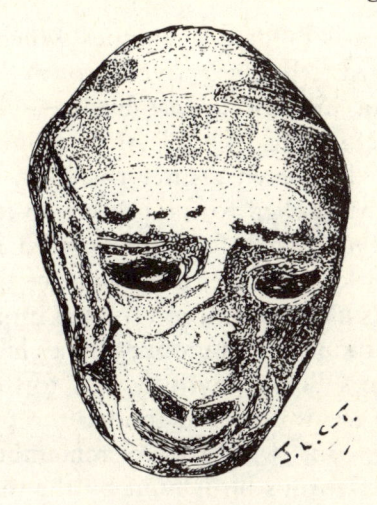

9. Pupae of two species of moth *(Spalgis)* which mimic the faces of monkeys.

wolf-snake, a bird-eating reptile widely distributed in the Oriental regions. Even more surprising is the extraordinary resemblance of some butterfly pupae of the genus *Spalgis* to monkeys' faces. One Oriental species is said to look like the common macaque of the region, while a related African species bears some similarity to the face of a chimpanzee. The difficulty in accepting the explanation of disguise lies in the great difference in size between a pupa and a monkey's face. A similar difficulty lies in explaining the crocodile-like appearance of certain Brazilian fulgorid bugs, commonly known as 'lantern-flies'. In many genera of these tropical insects, the head is greatly drawn out to form a huge, hollow, nose-like prolongation that was, at one time, erroneously believed to be luminous. The alligator-bug, in particular, may well gain protection from its disguise. It is not inconceivable that, just as a man may recoil in horror from a piece of rope that he mistakes for a snake, so may a monkey that gets a fleeting impression of an alligator be startled sufficiently to enable the insect to escape. Even if the bug is not mistaken for an alligator, or caiman, the appearance of a row of formidable teeth may, in itself, prove to be a deterrent to further investigation. Since other large bugs, such as cicadas, are relished by monkeys, there may well have been heavy selection in favour of an alligator-like appearance.

Tooth & Claw

If it is predatory enemies which, by avoiding such bugs, have proved to be a selective factor in the evolution of their disguise, it is not impossible that birds – which lack binocular vision and therefore are less able than monkeys to judge size – may have played an even greater role than monkeys. The same argument can be applied to the protective resemblance of a butterfly pupa to a monkey's face. In Howard Hinton's words: 'It seems too much to attribute to coincidence the fact that the Oriental species looks like a common Oriental monkey and the African one like a common African monkey. The chief argument against such resemblances being anything but fortuitous is the difference in size between mimic and model, the pupa being only 4.5 to 6.5 mm long. But we have to remember that a high proportion of insectivorous birds hunt by the method of "rapid peering"; they

10. *Above*, head of the alligator-bug *(Fulgora)* showing resemblance to a spectacled caiman *(Caiman), below*. Not to scale. (Drawn from photographs.)

peer at objects from several different angles in rapid succession. Their binocular field is probably usually so narrow as to be of little practical importance, and perception of solidity and distance is gained by evoking parallax. But with birds as with us and other vertebrates, the apparent distance of a familiar object is determined by the size of its image upon the retina. The method of rapid peering of insectivorous birds is going to mean that from time to time a bird will suddenly have a close-up frontal view of one of these pupae.'

Despite such positive statements, many people find it difficult or almost impossible to accept the argument that the superficial resemblance can be due to anything more than chance. Indeed, this is one of the many problems not yet resolved in relation to the appearance and coloration of animals. Another example, in which the selective factor could have been Man himself, is afforded by a Japanese crab, the pattern on whose back bears a striking resemblance to the face of one of the Samurai. Of this animal, Julian Huxley once wrote: 'The resemblance of *Dorippe* to an angry traditional Japanese warrior is far too specific and far too detailed to be merely accidental: it is a specific adaptation which can only have been brought about by means of natural selection operating over centuries of time, the crabs with more perfect resemblances have been less eaten.' For, although quite edible, these crabs are never harmed by Japanese fishermen. No more, for that matter, are the alligator-bugs of South America harmed by the local people who hold them in great dread, believing that the head is filled with a poison that kills within twenty minutes anyone who comes into contact with it. No doubt the uncanny resemblance to an alligator or caiman would be enough to engender superstitious dread among uneducated country folk. Likewise, it is easier to conceive a human being imagining a monkey's face in an insect chrysalis than to accept that a bird would do so.

Another ruse to deter predators is adopted by the pupae of one West African blue butterfly which are so covered with white hairs that they appear not only to be quite dead, but also to have been heavily infested by mould. Again, *Limnitis*, a member of the Nymphalidae (the largest family of butterflies) always pupates in a most exposed position on the upper side of a poplar leaf. Before pupating, however, the caterpillar attaches the stalk of the leaf to the twig so that it cannot fall to the ground if it should later become detached. The pupa is whitish-yellow with numerous black spots on its exposed parts. A swelling at the base of the abdomen is orange-yellow, shiny and semi-transparent: the cuticle behind is mat while in front it is highly polished. The appearance is such as to suggest that some animal, probably a bird, has attacked the pupa and discarded it as distasteful. Since birds and other vertebrates have to learn the significance of warning coloration by experience, as we shall see in a later chapter, in nature a small proportion of animals with conspicuous warning colours must be attacked and then discarded when they are found to be distasteful or poisonous! In yet a different case, the chrysalids of one small Brazilian butterfly arrange themselves in clusters of three, fastened by the tail along a stalk, so that the group resembles an inflorescence with seed pods or dried up buds.

At the beginning of this chapter, I mentioned some predatory animals that, through disguise, lure prey into their jaws. Equally, or even more common in the world of nature, are disguises that deflect attack either to a relatively unimportant part of the body or even away from the animal altogether. For instance, dummy eyes are found throughout a wide range of animals. In some instances they serve as bluff but, in others, they are believed to divert attack to non-vital parts of the body, thus increasing an animal's chances of escape. In some of the hairstreak butterflies, the wings are held vertically when at rest. A dark spot on the tip of the underside of the hind wing then looks like a head; antennae are represented by the 'tails' of the hind wings, and there are wavy marks on the wings which suggest legs. The illusion is heightened by movements of the tails in the slightest air current, in much the same way that the butterfly moves its antennae. A predator, such as a bird or lizard, might easily be deceived into attacking the less vulnerable wings rather than the head but, as in the case of so many of the phenomena recorded in these pages, there are no field observations or laboratory experiments to support the conclusions that have been inferred.

The appearance of two heads is by no means uncommon among reptiles. Wall-lizards, agamids, 'horned toads' and some geckos wriggle their tails when stalking prey, and a few snakes do the same. Similar behaviour has been recorded in young crocodiles. It attracts the attention of the prey to something, beyond the critical distance at which flight is engendered, for long enough to allow the predator to strike. In one American midget boa-constrictor and some coral-snakes, the tail resembles the head, but it is not clear whether the tail movements function to distract prey, deceive predators, or both. Perhaps equally deceptive are cases in which the tail does not mimic the head but is even more conspicuous, as in the kraits to be discussed later.

Amphisbaenians are a group of legless reptiles, like slow-worms, but whose affinities are uncertain. All known species are subterranean and live in tunnels burrowed underground from which they sometimes emerge at night. When disturbed, one large South American species raises both its rounded head and the stumpy tail, which can barely be distinguished from it. The vernacular name of this reptile is 'culebra de dos cabezas', meaning 'snake with two heads'. Whether the function of raising the tail is to attract the attention of potential predators away from the head is by no means certain. It may simply represent a means of confusing possible enemies as to which is the anterior end. It could also be a mechanism to give the impression that double the number of animals are actually present, or even, by simple mimicry, confer on the amphisbaenian the appearance of a rattle-snake.

Concealment is often almost impossible for a spider on an orb web because the opaque body of the spider presents a conspicuous contrast to its web, which may be almost invisible. Consequently, the spider is liable to attack from passing birds. This risk is reduced by spiders of the genus *Cyclosa* which build one or two false hubs of silk on their webs. At the centre of each of these is placed a bundle of scraps about the same size as the spider's own body so that, even if a bird knew that one of them was edible, the chances of it attacking the actual spider would be much reduced. In South America there is a related spider which places on its web pieces of bark and lichen, cut from a tree, which are of the same size and colour as its own body. Other web-builders decorate their snares with a bold zigzag or spiral of silk which catches the eye, distracting attention from the spider as well as reinforcing the strength of the web. Another effective device is to build the web across a dummy twig, made from scraps of débris, in which is left a space that fits the spider's body exactly. When alarmed, the spider darts into this space and vanishes entirely from view! Further examples of the use of disguise by orb-web spiders to deflect attack by predators will be given in Chapter 10.

Not only web-spiders, but many other kinds of animal, by clothing themselves with material from their surroundings are able to escape the notice of their enemies. Some South American spiders carry dead ants on their backs, as we shall see in Chapter 9. The resemblance of these spiders to ants is enhanced by the jerky, ant-like manner in which they run about. A looper caterpillar from Borneo fastens flower buds to the long spines of its body so that it comes to resemble an inflorescence; and the buds are renewed when they wither. The larva of the blotched emerald moth changes its disguise with the season: in summer, it fastens fragments of the leaves of its food plant onto its hooked bristles but, after hibernating for the winter, it uses scales or husks of oak. Of course, this may be entirely due to chance – when fresh leaves are not available it has to make do with something else!

Some soil-dwelling mites and beetles plaster mud and vegetable matter on their backs, and so become indistinguishable from their background. Crabs, and other kinds of animals often do the same, and usually exercise great care in just how the adventitious material is arranged on their bodies so that the best camouflage is obtained. Flatfishes not only change colour to resemble the background on which they are resting, but also scatter sand over their bodies so that little more than their eyes are exposed.

Animals sometimes disguise themselves by attaching living plants, or even other animals, to their bodies, a type of mutualistic association which is another example of commensalism. Many molluscs and crabs have seaweed, barnacles or hydroids growing on their backs. There is even a scorpion-fish in Indian waters whose camouflage consists of an hydroid attached to its body that apparently grows nowhere else in the sea. Further instances will be cited in Chapter 9. No less than three different subfamilies of weevils grow veritable gardens on their backs and legs. They are found in the high altitude moss forests of New Guinea. In these miniature gardens, there are various species of algae, fungi, lichens, mosses, and liverworts, sprouting from specially modified pits, surrounded by stout hairs on the bodies of the beetles. Furthermore, like real gardens, the miniature shrubberies are inhabited by a number of pests, including bacteria, pro-

Disguise

tozoa, nematode worms, rotifers or wheel-animalcules, and beetle-mites! This probably represents the ultimate in disguise, and it is not surprising to find it in a rain-forest, the most stable and complex of the world's terrestrial biomes, and one in which natural selection has been operating between competing species for countless millions of years because the physical conditions of life have changed so little.

The use of adventitious materials, living or non-living, for the purpose of disguise, overcomes a difficulty that few animals have surmounted – that of escaping from their natural symmetry. Thus, the wing markings of cryptic moths that resemble lichen are invariably symmetrical, as are the bodies of stick- and leaf-insects. Why this should be so is not known. One has only to look at a snail with its spiral shell, or at the body of a hermit-crab or a flatfish, to realize that animals are by no means invariably symmetrical: yet asymmetry in protective resemblance is apparently never found. When protective resemblances can be so extremely detailed, there would seem to be no inherent reason for the absence of asymmetry in disguise. Possibly the advantages would not justify the cost in terms of the additional genetic complexity that would be required. As always in nature, economy is of paramount importance.

4 Armour

The design of a tank represents a compromise between the conflicting requirements of speed, armour and fire-power. In general, battle tanks with massive armour mount big guns and are used for infantry support or to destroy the armoured fighting vehicles of the enemy; but they lack speed. Light, medium or cruiser tanks are faster but less heavily armoured, and often carry less formidable armament. Their roles in warfare lie in reconnaissance or to exploit a break-through. The same relationship between speed and armour may be discerned among animals. Woodlice, for example, can be divided into two kinds. Pill-woodlice *(Armadillidium)* move slowly but conglobate, or roll up into a ball for defence, relying on their tough integuments to withstand the fangs of hungry spiders and centipedes. Garden slaters of the genera *Oniscus* and *Porcellio*, on the other hand, have longer legs and are faster, but their exoskeletons are relatively soft and thin. At the same time, however, they produce chemical secretions that make them distasteful to most spiders – although a few species of hunting spiders of the genus *Dysdera* actually specialize in eating them!

Neither armour, speed nor the repellent chemicals of woodlice are very effective against their larger vertebrate enemies such as birds and toads, but these are normally avoided through the secretive habits and nocturnal behaviour of woodlice in their daily lives; and although pill-woodlice, like pill-millipedes *(Glomeris)* tend to wander into the open more often, they are so big that the tiny mouths of shrews cannot bite them, and their shiny integuments make them too slippery for birds and other larger predators to grasp at all easily. Furthermore, the diameters of some woodlouse species are increased by the presence of numerous tubercles on the integument.

Armour

The surface of the body of most animals is protected by the skin, which may be surprisingly tough, but additional armour is often provided by calcareous skeletons, shells, scales or hard integuments, like those of woodlice, crabs and lobsters. Corals are colonial animals belonging to two different classes of coelenterates. Both kinds secrete strong, calcareous, skeletal structures within which the microscopic individual feeding polyps can retreat. Milleporine corals are related to the freshwater hydras. Common members of the faunas of reefs, they are able to sting through the human skin, causing painful weals. Most large corals, however, belong to another group, which contains the sea-anemones, whose stinging cells cannot penetrate the skin of man. It also includes many reef-building species, such as the brain-corals *(Meandrina)* and mushroom-corals *(Fungia)*, the 'organ-pipe' coral *(Tubipora)*, the precious coral of commerce *(Corallium)*, as well as 'dead man's fingers' and other soft corals.

Despite their protective skeletons and stinging cells, corals are by no means safe from attack. They are nibbled away by parrot-fishes, boring bivalves, worms and sponges, while the crown-of-thorns starfish *(Acanthaster)* sucks the polyps from their calcareous shelters and destroys large areas of reef. These spectacular starfishes which, in recent years, have been threatening much of the Great Barrier Reef, grow to nearly a metre in diameter,

11. Hunting spider *(Dysdera)* eating a woodlouse. (After W. S. Bristowe.)

and often have more than a dozen arms. They feed by everting the stomach, spreading its filmy gastric membranes over the coral and applying them closely. The flesh of the coral polyps, including the stinging cells, is then liquefied by digestive juices and absorbed. When the starfish moves on, it leaves only the dead white skeleton of the coral behind, and this is quickly broken up and washed away by the waves.

In addition to affording protection from predatory enemies, calcareous skeletons lend structural support to the corals that secrete them. The same is true of the tubes of fan-worms and their allies. Apart from making their possessors somewhat distasteful, however, the spicules of sponges and the internal skeletons of radiolarians and foraminiferans – countless millions of which form the chalk deposits of the downs – probably serve little function other than one of providing support, and are not really defensive. Whereas foraminiferan skeletons are constructed of calcium carbonate, those of radiolarians are usually siliceous and composed of quartz. Microscopic in size, both are extremely beautiful when magnified. Accumulations of radiolarian shells are less plentiful than chalk deposits, but the island of Barbados is said to be composed largely of them although its soil is volcanic in origin.

Moss-animalcules also form sedentary colonies, the individuals of which are protected by a horny, calcified investment. Thus, the colonial sea-mat *(Flustra)* has numerous polyps lying in a protective skeleton which is branched to form fronds that look like pieces of seaweed. Each of the chambers in which the polyps live is guarded by a lid which, when open, allows the animal's tentacles to protrude so that it is able to feed. When danger threatens, however, the polyp withdraws into its investment and closes the lid. The partial calcification of the skeleton allows it to remain flexible so that it can bend with the movements of the water and, consequently is not torn from its moorings on rocks by the force of the waves.

Perhaps the most efficient armour in the animal kingdom is to be seen in the mollusc shell. Each kind of mollusc has its own characteristic type of shell. The chitons have shells usually of eight pieces or valves, held together by a surrounding fleshy girdle; monoplacophorans have one-piece, limpet-like shells; tusk shells are tubular and tapering at both ends; while the bivalves have two-piece shells, hinged at one edge and equal or unequal in size and shape. Gastropoda, such as limpets, snails and whelks, have univalvular shells, usually coiled and external, but slugs, sea-hares and so on are without shells, having lost them in the course of evolution. Finally, the Cephalopoda – nautiluses, octopuses, cuttlefish and squids – may be without shells but, more frequently, have shells which are reduced in size, coiled or straight, and usually internal. Such shells have lost their protective function: they have become devices for regulating the buoyancy of pelagic, swimming forms.

A shelled mollusc grows by adding calcareous material to the edge of the tiny valves or embryonic coiled shell acquired during the larval stage. Many species produce shells which differ conspicuously at different stages of growth: the shells of others are essentially similar throughout development and have an incomplete appearance, even at maturity. Crystals of calcium carbonate are deposited in layers by the fleshy mantle of the growing mollusc, either as aragonite – the mother-of-pearl or 'nacre' which lines the inside of many shells – or as calcite, a less dense and smooth form of calcium carbonate. A horny substance, known as 'conchiolin', is associated with the layers of calcium carbonate and helps to strengthen them. It also forms a brownish outer covering or 'periostracum' which affords protection against carbonic acid in the water, which could otherwise dissolve or damage the main structure of the shell. In nature, the beauty of many shells remains hidden until their owners have died and the periostracum has been worn away by the action of waves and sand. Not all shells have a periostracum, however. Cowries, for instance, have shiny, glossy shells, the outsides of which are surrounded and protected by the tissues of their mantles.

Not even the strongest shell provides absolute protection from predatory animals, of course, but the shells of aquatic molluscs, whose weight is partly supported by the water, tend to be thicker and stouter than those of terrestrial snails or of swimming species which must be light in weight if they are to float. A beautiful example of the latter is the purple sea-snail *(Ianthina)*, whose camouflage was discussed earlier, and whose frail, violet-tinted shells are so often tossed intact upon tropical shores. The purple sea-snail spends its life afloat, suspended upside down from a raft of bubbles to which some species attach their egg capsules. The necessary reduction in weight, however, has had to be achieved at the expense of the shell, which is thin and weak, and the animal relies on camouflage for defence.

The efficiency of marine shells in giving protection to their occupants is reflected in their use by hermit-crabs, after the death of the original owners. Hermit-crabs have soft, asymmetrical abdomens with terminal sickle-shaped appendages which grip the central pillars of the shells in which they live. That they are so common and well known is clear evidence of the success they have achieved through sacrificing much of their own protective armour so that they can obtain shelter within the empty shells of univalve molluscs. Ungainly in movement and burdened by the weight of the acquired shell, they are admirably adapted for life on rocky shores churned by powerful seas and exposed to predatory sea-birds at low tide. Even the claws are asymmetrical, that on the right being larger, flattened, and more heavily armoured so that it can be used to close the aperture of the shell when the hermit-crab withdraws. Hermit-crabs are also very numerous in deeper waters where their thick shells save them from being eaten by fishes, octopuses and lobsters, which could easily swallow normal crabs of comparable size.

Many shore molluscs, such as limpets, attach themselves to rocks with a strong adhesive grip. Others, such as the dog-whelk *(Nucella)*, easily lose their grip, but have shells that are so robust that they permit much buffeting by the waves. The entrance to the shell can be closed by a circular plate or 'operculum', but this is seldom employed, except when the animal loses its grip on the rock. Dog-whelks prey on barnacles, as well as on mussels and other molluscs. Not every predator is able to feed on such animals, encased as they are in stout shells and so firmly attached that they can resist the full force of the sea, but dog-whelks attack them insidiously. They settle on their prey and bore a small hole through its shell with the narrow ribbon of teeth or 'radula' carried at the end of the whelk's proboscis. This is a purely mechanical process, but there are other predatory marine snails which aid the process with acid which dissolves the chalky shell on the prey. Once the shell has been perforated, the proboscis is extended into the tissues of the prey. These are rasped out by the radula teeth and passed back, as on a conveyor belt, into the mouth of the dog-whelk.

Whelks are by no means the only enemies of mussels. Starfish grip the two valves of the shell with the suctorial tube-feet that are clustered along the under-surfaces of their five arms. With a steady, but continuous pull they are eventually able to overcome the greater strength of the adductor muscles of the bivalve, whose shell therefore gapes open. The starfish then protrudes its large stomach and digests the mussel externally.

In the Russian Arctic tidal zones, mussels often make up more than 70 per cent of the total animal weight per square metre and, in Britain, mussel populations can reach a density as high as 140–170,000 per square metre. It has been found that Danish mussels, although belonging to the same species as the English variety, are up to four times stronger. They close their shells more tightly, are more difficult for starfish to open and accordingly, it is said, are preyed upon less by them. This may be so, but it would seem more probable that the extra strength may have been the result of even heavier predation from which only the toughest mussels escape to reproduce themselves.

Starfishes are inveterate enemies of all small bivalves, even eating scallops and their relatives, the queens – if they succeed in catching them. This is by no means an easy task, however, because scallops and queens have forsaken the sedentary habits of their distant ancestors and are very active animals. They swim by opening and closing the valves of the shell: this expels jets of water which drive them jerkily through the sea. The place where the water is forced out depends upon the disposition of curtains of tissue which can be extended or withdrawn locally. In normal swimming, the jets are emitted near the hinge so that the free edge of the shell goes in front and the scallop appears to be swallowing water in a series of gulps.

Scallops and queens are able to detect changes in light intensity, and even perceive moving objects with the aid of a row of eyes along the edge of the mantle where the shell opens. Consequently, they are alerted when anything approaches them. If a starfish appears they recognize its distinctive smell and immediately carry out what is known as an 'escape reaction'. The adductor muscle contracts instantaneously, while the curtains around the free margins of the shell are drawn back so that a jet of water is expelled between them. This rapidly propels the mollusc away from its enemy, hinge foremost.

Molluscs are not the only animals to possess valves. Lampshells, or brachiopods, are also protected by a pair of shells which appear superficially to be like those of bivalves. They differ, however, in that they have upper and lower valves instead of one valve on the left and another on the right side of the body, as in bivalve molluscs. Brachiopods live on the sea-bottom, mostly in deep water, and are nearly always fixed by a stalk. They are not particularly numerous at the present day, but the group is a very old one which flourished in byegone ages. Perhaps the lampshells suffered from competition with bivalve molluscs, which beat them at their own game!

Since molluscs are highly palatable, and form an important item of food for many kinds of bird, it is not surprising to find that many snails are camouflaged or disguised in one way or another. A species that has been intensively studied is the terrestrial banded snail *(Cepaea)*, widely distributed throughout western Europe. The ground colour of its shell is yellow, pink or brown. The shell may be unbanded, or with any number up to five dark bands running round the whorls. The frequencies of the various shell colours vary in different habitats as a consequence of predation by thrushes. In some areas, the song thrush eats many banded snails, particularly in dry weather when other food is not available. Since this bird breaks snail shells on particular stones or 'anvils', it is easy to determine the number of snails of different colours that have been eaten, through regular collection of broken shells from thrush anvils. In one woodland area of Britain, over 40 per cent of shells found at thrush anvils were yellow, although a random collection showed that yellow snails formed only 24 per cent of the total population. In April, the ground was mainly brown with little vegetation so that yellow-shelled snails were the most conspicuous. By late May, however, when the backgound had become predominantly green, the proportion of yellow shells found at song thrush anvils dropped to 15 per cent and mainly darker snails were being devoured.

It seems remarkable that slugs, which are related to land snails, should have lost their shells in the course of evolution. Presumably they are to some extent protected by their thick slimy skins, but they must surely have gained something in recompense for losing their shells, otherwise their naked condition would never have evolved. Possibly they are better able to creep through dense vegetation without a bulky shell.

Marine shell-less molluscs are not faced with the respiratory, mechanical and sanitary problems that result from the restrictions imposed by living in a shell, but they cannot afford to give up their armour unless they acquire other defensive adaptations to protect them from predators. Many sea-slugs, for example, are cryptically coloured. If detected and attacked, however, they erect and wave the numerous defensive papillae or 'cerata' with which the body is covered. These are armed with the stinging cells of the sea-anemones on which sea-slugs feed, and any enemy touching them is likely to be stung, as described in Chapter 8. In addition, they possess repugnatorial glands and, if the predator is still not deterred, the cerata are readily autotomized or broken off, to be regenerated subsequently. Meanwhile, the mollusc makes its escape while the predator is enjoying its unattractive meal.

Most cephalopods have internal shells which give no protection to their owners, but exceptions occur in the nautiloids whose spiral shells are divided into a number of chambers. The animal lives protected by the outermost of these, its tentacles projecting from the shell to seize the small animals on which it feeds. The pearly nautilus of Indo-Pacific waters was at one time thought to be rare because it only comes to the surface of the sea at night, when water is pumped out of the chambers of the shell and replaced by air which provides buoyancy. During the day, the nautilus allows its shell to fill with water, and sinks to the bottom of the sea where it cannot be seen. The 'bone' of the cuttlefish also serves as a reservoir of gas which helps the animal to float by reducing its density but, since it is internal, it cannot provide shelter from enemies. Like squids and octopods, the cuttlefish relies for safety not only on its ability to change colour with extreme rapidity but, also, on the fact that it can swim very quickly by expelling jets of water from its mantle cavity through a ventral tubular funnel. The funnel is highly mobile and can be pointed in any direction to control movement. Squids are streamlined with tapered bodies: unhampered by an external shell, they are extremely speedy and have even been known to leap out of the sea on to the decks of ships, four metres above the surface of the water. They can also move almost imperceptibly when approaching their prey.

Tooth & Claw

Another type of armour takes the form of scales. Vertebrate scales occur only in fishes, reptiles, and on the legs of birds; but scale-like structures are found in some mammals – the pangolins or scaly ant-eaters. They cover the whole body and are formed of consolidated hairs. To such spiny scales, coupled with the secretion of a nauseous smell, the pangolins must owe their survival – for they are comparatively inactive and unintelligent. When attacked by a fox or jackal, they seldom try to escape but, like hedgehogs, roll into a compact ball presenting to the enemy an impenetrable armour of horny plates while emitting, at the same time, their obnoxious odour.

Although scales are primarily protective they may, in certain circumstances, acquire secondary functions. They assist, for instance, in thermal regulation of the scaly lizards *(Sceloporus)* of North America. Minute projections from the tips of the scales help to shade the bodies of the lizards when they are facing the sun. Indeed, the suggestion has been made that feathers may originally have evolved from such scales, and that their initial function was possibly one of insulating the body.

In coffer- or trunk-fishes *(Tetrosomus)* the scales are joined edge to edge to form a ridged case from one end of which projects the mouth, from the other the naked tail. The case can have three, four or even five sides, and one or more of its edges may be armed with strong spines. The little pine-cone fish *(Monocentrus)*, from the coast of Japan, although belonging to a totally different order, is another form in which thick scales unite to enclose the body in a box. During growth, the box enlarges by addition to the edges of the scales. In cases such as this, it is obvious that mobility has been sacrificed for armour.

The heaviest armour in the animal kingdom is to be found among the arthropods and vertebrates; but sea-urchins, starfishes and sea-lilies are also protected by hard exoskeletons. These are composed of limy plates, called 'ossicles', embedded in the skin. Especially in starfishes and brittle-stars, this retains some flexibility while, at the same time, forming a hard layer not easily penetrated by the teeth of fishes or the claws of lobsters and crabs.

The exoskeleton or cuticle of arthropods consists of a number of layers secreted by the underlying epidermal cells. The inner layer is composed of a tough, resilient, cellulose-like polysaccharide known as 'chitin' which forms a fibrous framework

Armour

that is strengthened by other materials. The chitin of crustaceans, whose bodies are usually supported by water, is reinforced by chalky substances. Apart from woodlice and millipedes, the integuments of which are calcified, the terrestrial arthropods rely upon lightweight armour composed of chitin impregnated with

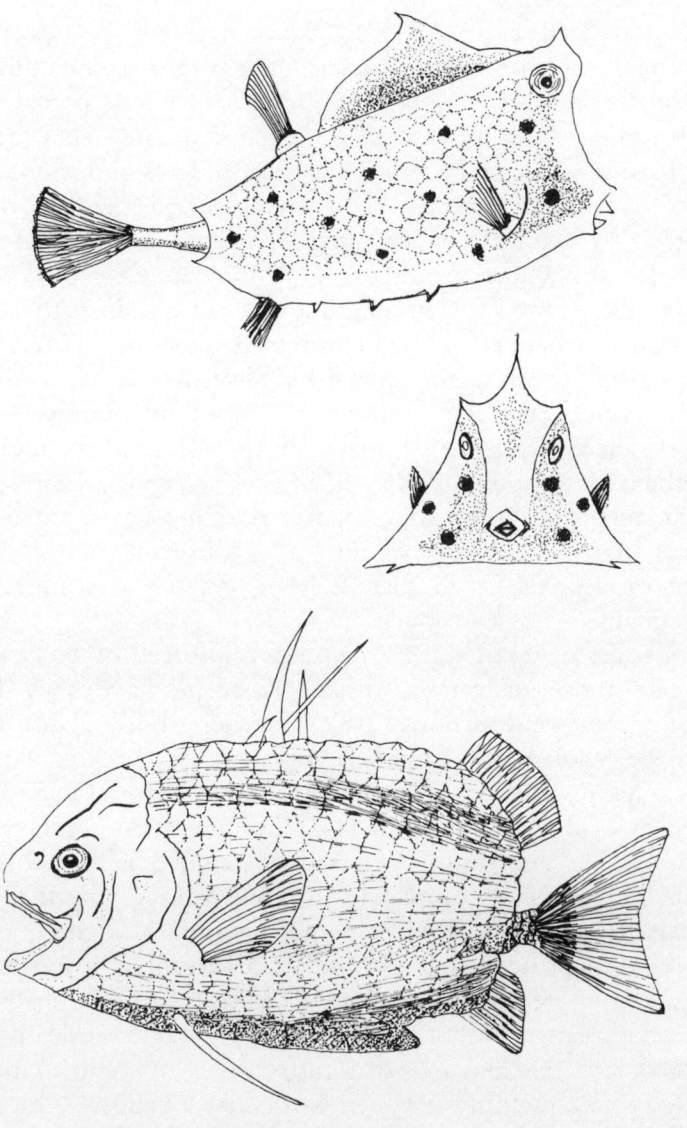

12. *Above*, coffer-fish or trunk-fish *(Tetrosomus)*; *below*, pine-cone fish *(Monocentrus)*. (Modified after P. H. Greenwood.)

'sclerotin'. This is secreted by the epidermal cells in the form of a protein that is subsequently 'tanned' by a quinone. Just as leather becomes resilient, tough and impervious when it is tanned, so the cuticles of insects and arachnids are hardened and made comparatively waterproof when they are sclerotized.

Especially in beetles and certain mites, the cuticle may be extremely hard and resistant. Adult darkling beetles (Tenebrionidae) – characteristic of desert and savanna regions throughout the world – are so tough that few invertebrate predators are capable of crushing their cuticles. At the same time, they are protected from vertebrate enemies, such as rodents and mongooses, by the nauseous chemicals they secrete. Indeed, these beetles are ignored by almost every animal except the camel-spider *(Galeodes)* for whose formidable jaws they are clearly no match. With bodies some 7 cm in length and legs that can span a width of 15 cm, camel-spiders are impressive creatures. Their jaws, possibly the largest in the animal kingdom in relation to the size of the owner, take the form of two powerful pincers. Camel-spiders emerge from their burrows in the silence of the night and glut themselves upon a variety of insects, scorpions, lizards, mice and small birds. When they catch a darkling beetle, their meal sounds like the cracking of nuts at a Christmas party! In the world of animals, as in that of Man, defence is almost never impregnable.

The scales of reptiles are sometimes reinforced by bony plates, just like those of certain fishes. Tortoises have very heavy armour, consisting of horny scales overlying bony plates, which cover the whole body. The shell of the animal's back is known as the 'carapace', the part underneath as the 'plastron'. The two are joined by bridges between the front and hind legs on each side of the body. The plastron of American box-turtles is hinged so that the shell can be shut completely. Anyone picking up one of these animals has to be careful not to get his or her fingers pinched! This kind of protection does have its disadvantages, and breathing is complicated for an animal encased in a box-like shell. Turtles and tortoises cannot expand their rib cages: instead, air is pumped into the lungs by the movements of internal muscles inserted into a membrane which works like a bellows. One set of muscles widens the body cavity so that the lungs are expanded and air drawn in. Another set of muscles then presses the viscera,

Armour

or internal organs against the lungs, causing the air to be expelled. These actions are accompanied by movements of the head and forelimbs.

Although tortoises can sometimes be scooped out of their shells by large predators, such as jaguars or leopards, they are usually safe from the attacks of smaller carnivores once they have grown old enough for their shells to have become fully hardened. As suggested earlier, it seems probable that full-grown giant tortoises die only as the result of disease, accident or old age – a sublime condition rarely found in the animal kingdom.

Most animals which cannot quickly take flight are vulnerable to the attacks of marauding columns of driver- and safari-ants. The African python, sleeping after a heavy meal, will be reduced to a polished skeleton within a few hours. Not many kinds of tortoise are to be found in rain-forest inhabited by driver-ants; but Bell's box-tortoise *(Kinixys)* is practically immune to them since its shell and the tough scales of its forelegs, when drawn in, protect the vulnerable head almost completely.

A number of lizards have their scales thickened to form armour but, because they are relatively small, they are easily snapped up by large predatory birds and mammals: so they can

13. Armoured dinosaur *(Triceratops)*. (Reconstruction.)

Tooth & Claw

never afford to relax their vigilance while basking in the sun. In another group of reptiles, the crocodiles and alligators, the scales of the back are thickened into armour. Despite this, jaguars, leopards and lions have been known to kill big caimans and crocodiles although, under somewhat different circumstances, they might themselves have fallen victim to these reptiles. Quite large crocodiles are sometimes killed by hippos, or are trampled to death by the elephants which they have been sufficiently unwise to attack. Dragged ashore and squashed, the unfortunate reptiles are then not infrequently lodged in the branches of a tree. Why elephants should do this is not known, but several instances have been recorded.

Among extinct reptiles, the greatest development of armour was found in the dinosaurs. *Triceratops*, for instance, wore a heavily armoured collar as well as three sharp horns on its head; while *Scolosaurus* had heavy rows of scales produced into conical shapes on its head, back and tail. The most advanced of the armoured dinosaurs were the nodosaurs and ankylosaurs. Their bodies were protected by a mosaic of heavy, pavement-like plates of bone. In life, these were probably covered with horny scales, like the shell of a tortoise. The skull, too, was almost buried in a helmet of thick, dermal bones, in which the tiny eyes were deeply sunken. Their small teeth show that the nodosaurs must have fed on soft vegetable matter, while their defence depended entirely upon the heavy armour. Like enormous armadillos, these harmless reptiles grazed peacefully unless threatened by a giant meat-eater such as *Gorgosaurus* or *Tyrannosaurus*. Then they would almost certainly have crouched down on the ground, presenting their foe with a solid defence of heavy armour. They were probably extremely difficult to overturn, should any predator have attempted to reach the vulnerable underparts. Moreover, this would have been a dangerous manoeuvre for the heavy, spiked tail was an effective weapon, capable of inflicting deep and painful wounds on the adversary.

Some of the nodosaurs had tails shaped like the maces and flails that bishops and other churchmen customarily wielded in medieval days when they felt the urge to join in battle. For some unknown reason, it was considered more appropriate for holy men to crush the skulls and chests of their opponents with iron clubs rather than to pierce them with spears and arrows! With a

14. Pangolins curl up in defence.

fearsome combination of mace and armour, the nodosaurs survived for several millions of years until they were finally caught up in the great wave of extinction that spelled the doom of the ruling reptiles.

No mammal is armoured as heavily as some of the dinosaurs were, but the rhinoceros has a skin so hard and tough that it was at one time used for making shields which were impervious to the heaviest spears that could easily have penetrated wood of equal thickness. Even more effective is the bony armour of the armadillo, which is interspersed with areas of softer skin. These act as hinges and enable the animal to roll into a ball, so protecting its soft belly.

When threatened by danger, the different species of armadillo react according to the shape of their armour plating. In some *(Tolypeutis)*, the shield overlying the shoulder blades is shaped like a horseshoe in section so that the head and tail fit together in a remarkable manner when the animal rolls into a ball. The armour is made up of broad, horny scales which overlap one another like tiles on a roof. The shield of *Euphractus*, on the other hand, is flattened on a low arc so, if taken by surprise and unable to escape into its hole, this armadillo can do no more than press itself firmly against the ground. Another armadillo *(Chlamyphorus)* uses its posterior shield to close the burrow in which it takes refuge. The female pangolin even shelters her young, in case of danger, by placing the baby on her belly and then curling round and protecting both it and herself with her long scaly tail.

It would be useless for an animal to develop armour or, for that matter, any other defensive structure, unless it were simultaneously to evolve appropriate behaviour patterns. Which came first, the armour, or the habit of curling into a ball; warning coloration or the behaviour pattern leading to its display? Of course, to a large extent, both have developed together, but the very first evolutionary stirrings must have been of a behavioural nature, rather than modifications of form and shape. Perhaps the ancestors of the armadillos just happened to have rather tough hairs and skin, and defended themselves by curling up. Gradually, they evolved armour which enhanced the efficiency of this reaction to danger. At the same time and in consequence, the need to move quickly was reduced and the animals specialized on armoured defence. It would be uneconomical and impractical to become both speedy *and* armoured.

5 Barbs and Spines

Parallel evolution has taken place over and over again in the animal kingdom, and this book is replete with examples. Each chapter reveals one case after another of some simple principle of defence that has been exploited by innumerable different animals living in different environments.

Protected by their armour plates, armadillos curl up when attacked by enemies; and animals with defensive spines do the same. A hedgehog rolls itself into a ball and erects its quills when it is frightened. This action makes it very difficult for a predator to attack. (Nevertheless, the hedgehog does form part of the diet of weasels, badgers, vultures and the larger owls. The fox is reputed sometimes to push a hedgehog over a bank and, while it is bewildered by the fall and consequently not tightly rolled, to bite its unprotected underside and kill it.) The quills of porcupines inhabiting the tropical and subtropical regions of the Old World form an even more efficient means of defence. If a porcupine is irritated, it erects its quills, grunts and puffs loudly, and produces a rustling sound by banging its spines together. When an aggressor approaches, the porcupine darts suddenly backwards and impales the enemy with its spines, sometimes inflicting serious wounds. In one recorded instance, a leopard was mortally wounded by a porcupine; in another, an almost adult tiger, its liver and lungs perforated in many places, was found dead a few yards from the body of its intended victim.

The American tree-porcupine may be less dangerous as its quills are shorter but, if a dog approaches too closely, it erects its spines, raises its back, lifts its tail, and bends its nose down under its body. Keeping its rear always turned towards the assailant, it strikes repeatedly with its tail, leaving some quills in the flesh of the aggressor whenever contact is made. These cause nasty septic wounds. Barbs on the tips of the quills cause them to work into the body of the victim with every muscular movement. In one case, a fragment about 1.9 cm in length moved nearly 5 cm in 30 hours through human muscle, while another only 0.6 cm long advanced almost 4.5 cm in less than two days.

The tenrecs of Madagascar are insectivores, like the hedgehogs of the Old World and, also like hedgehogs, rely for defence upon stiff spines on their backs and sides. The echidna or spiny ant-eater of Australia, on the other hand, is an egg-laying monotreme related to the platypus but quite unrelated to porcupines and hedgehogs. Nevertheless, it also has many sharp spines and rolls into a ball in the face of danger. The thylacine or marsupial wolf, now probably extinct, was the only predator that could easily kill an echidna. Hedgehogs, tenrecs, echidnas and porcupines provide a good example of parallel evolution – the acquisition quite independently by unrelated animals of similar mechanisms that perform the same function.

Unlike horns, teeth, claws and pincers, which are used not only in defence but also for fighting and the capture of prey, armour and spines have apparently evolved purely as anti-predator devices. Nevertheless, they appear in a considerable variety of shapes and sizes, and differ according to the predators that they are intended to deter. They may be long or short, sparse or clustered, smooth or barbed. Sometimes they are used in association with poison glands, and sometimes their owners display conspicuous warning coloration or make threatening noises when disturbed.

Defensive hairs and spines are not restricted to hedgehogs, porcupines and other mammals. The integument or cuticle of arthropods sometimes bears a dense covering of detachable hairs or scales which can serve effectively for protection. The minute, bristly millipedes *(Polyxenus)*, often to be found under lichen on tree trunks and stone walls, are covered with tufts of peculiar hairs shaped like small pin cushions. These hairs are hollow, easily detached from the body, and bear a number of barbs so that they become embedded in the jaws of any small insect or spider rash enough to bite so unprepossessing an animal.

The larvae of carpet-beetles also bear prominent tufts of segmented, spear-headed hairs on their abdomens. When attacked, they swing their bodies towards the predator, into whose body

Barbs and Spines

masses of barbs become embedded. Small enemies, such as ants and beetles, are effectively repelled, becoming hopelessly entangled, but toads, lizards, birds and rodents, of course, are too big to be deterred by such small hairs. Woolly-bear capterpillars, which are larger, are similarly avoided by most insectivorous birds, but their numerous urticating hairs do not protect them from cuckoos which actively seek them out. The hairs become embedded in the cuckoo's gizzard so that it appears to be lined with dense fur. This diet is an inherited rather than an acquired taste, which develops once the young cuckoo has left the care of its foster-parents – no matter what food it has previously received from them. This is yet another example of a defence mechanism which is generally effective except against specialized enemies.

It has been known, since the time of Pliny, that the urticating hairs of certain lepidopterous caterpillars are not merely effective defences against other arthropods, but can also cause cutaneous reactions in Man and larger animals. The effect of these hairs is not simply mechanical, but is dependent on a poisonous secretion inside them. Contact with urticating hairs induces dermatitis and allergic symptoms in Man, while their toxic nature is also indicated by the fact that the upper respiratory tract of the victim is sometimes affected. In some instances, the larval cuticle, with its covering of defensive hairs, may be retained as a protection by the pupal stage or chrysalis of the insect.

Larvae of the common European gold-tail *(Euproctis)*, one of the tussock-moths, are provided with urticating hairs composed of barbed spicules. Whether the irritating properties of these spicules are mechanical only, or partly due to a poisonous secretion, is not known. When she emerges from her cocoon, which is covered with, and protected by, larval spicules, the female moth collects some of them by brushing the cocoon with her anal tuft. These make her distasteful to enemies. Later, she distributes urticating hairs over the mass of eggs that she lays. In this way, all stages of development are protected by the barbed spicules produced by the caterpillars.

The spines and bristles of insects sometimes take the form of flattened scales. Whereas smooth-bodied insects that fly into the webs of orb-weaving spiders remain stuck there, moths merely lose some of their detachable scales to the viscid threads, and then fly on to safety. Other insects, such as adult caddis-flies, which

Tooth & Claw

are covered with loose hairs, and white-flies (Aleyrodidae) which secrete a waxy powder, may similarly be protected against the snares of spiders. In the same way, moths that settle on sundew plants – so lethal to most small insects – are often able to escape with the loss of only a few unimportant scales. Silver-fish, having tapered bodies covered with waxy scales, also escape from sundew plants. They can easily elude ants, too – slipping away and leaving the enemy with a mouthful of scales! The larvae of blue butterflies inhabit the nests of ants: they bear an especially thick covering of scales, which protects them in the same way, should they be attacked.

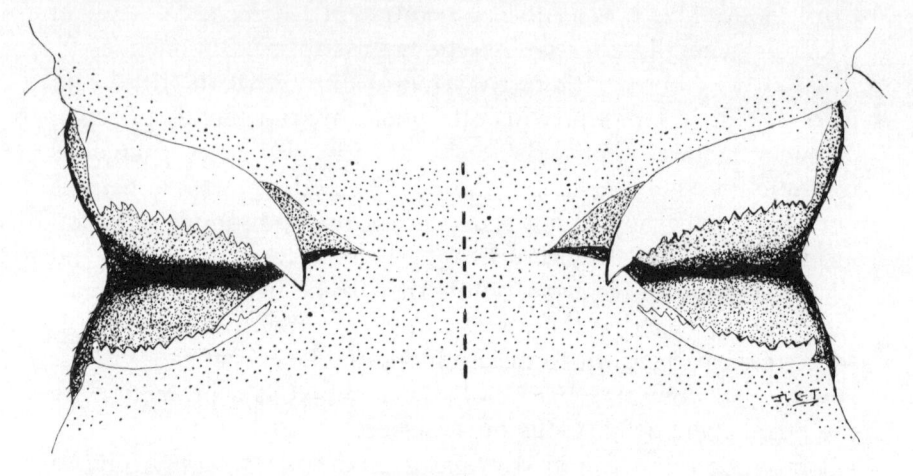

15. Second abdominal gin-traps of a beetle pupa *(Alphitobius)*. (Modified after H. E. Hinton.)

The pupae of many beetles, and those of some Lepidoptera, are equipped with spiny defensive organs, known as 'gin-traps' because they work like old fashioned rabbit snares. As we have already seen, these afford no protection from large vertebrate predators: the gin-trap is essentially a means of defence against mites and other animals much smaller than the pupae. The circumference of insect abdominal segments is less where they are joined to the adjacent segment in front or behind than in the middle. In consequence, the region near the posterior margin of one segment and that near the anterior edge of the segment immediately behind is always depressed. The evolution of a gin-trap needs no more alteration than a slight deepening of this

Barbs and Spines

depression, a hardening of the cuticle at the edge, and its extension into a keel-shaped, spiny blade.

Gin-traps are confined to the abdomen, the only part of a chrysalis or pupa that can be moved freely. At rest, the abdomen curves downwards so that the jaws of each dorsal gin-trap are held widely open. When the pupa is stimulated, however, the abdomen is bent sharply upwards, and the segments are partly telescoped into each other so that the jaws of the trap meet or overlap. Gin-traps on the sides of the body, on the other hand, are snapped shut when the abdomen makes lateral or eliptical movements. Gin-traps only remain closed for a fraction of a second – but pinch for long enough to cause an attacker to leave the pupa alone. If they were to remain shut for any length of time, the struggles of the enemy in its efforts to escape might well cause damage to the pupa. With the development of gin-traps we come to one of the frontiers between passive and active defence. Spines, like armour, are passive, but gin-traps are only employed in the presence of an enemy and are therefore active defence mechanisms.

The workers of many species of ants bristle with spines and tubercles which guard them from predatory beetles and other enemies. It is not uncommon to find at least a single pair of spines, on the dorsal side of the posterior thoracic segment, whose function is apparently to shield the vulnerable waist from the mandibles of a predator. Other ant species conceal their sensitive antennae in deep grooves, or under broad projecting ridges along the sides of the head. Since ants can also bite, sting, and employ various forms of chemical warfare, it is not surprising to find that their predators are not only rather few in number, but have developed special ways of catching and eating them.

Ants form the whole or part of the food of a handful of spiders, wasps and beetles: in some cases these predators also mimic their prey. Other insects, only predatory in their larval stages, construct pits in fine sand for the capture of ants. They include ant-lions *(Myrmeleon)*, the 'demons of the dust', and the larvae of certain flies *(Lampromyia* and *Vermileo)* which have independently adopted the same mode of life. At the bottom of their conical pits, these larvae lie buried, waiting for passing ants and other insects that may slip into the trap. The prey is assisted in the slide to its doom by sand particles flicked with surprising

Tooth & Claw

force and accuracy by the larva, lurking in ambush below, whose immense mandibles enable it to destroy its prey without the unfortunate ant being able to retaliate or come to grips with the soft abdomen of the aggressor.

Among vertebrates, several species of toads, lizards and snakes, as well as birds such as thrushes and woodpeckers, regularly feed on ants while others, including gulls and swallows, devour immense numbers of winged ants during their nuptial flights. Of the mammals, ant-eaters and ant-bears, as well as echidnas, have specialized on a diet of ants. Toads and lizards tend to be rather unselective, feeding on a variety of insects and other small animals. Worm-snakes are subterranean, and specialize on a diet of ants and termites, although they also eat other soft-bodied insects, earthworms and slugs. Their smooth, scaly bodies are invulnerable to the bites of the smaller species of ants on which they usually prey. Birds merely peck ants with their long beaks and keep their bodies out of range of their prey's defences.

Insectivorous habits have been secondarily acquired by some carnivorous mammals that have forsaken a diet of flesh. One of them, the aardwolf, to which we shall return later, resembles a hyena but is smaller and more slender, with weak jaws and simple teeth. It feeds mainly on termites and carrion. Reduction of the teeth is characteristic of ant-eating mammals – reaching an extreme in the toothless ant-eaters, the aardvark or 'ant-bear', pangolins, and the marsupial ant-eater *(Myrmecobius)*. These distantly related animals all have powerful claws which enable them to tear ant and termite nests apart, and long sticky tongues to explore the galleries of the colonies. When insects stick to them they are licked up. Physical resemblances between ant-eating mammals are a consequence of eating the same diet, and have been evolved independently in the various groups – another example of parallel evolution.

Some insects, as we have seen, derive protection from detachable coverings of an artificial kind. Caddis larvae build cases of sand, pebbles or twigs, inside which they live. Their houses not only make them inconspicuous on the bed of a pond or stream but also render them most unpalatable. They are analagous to a covering of spines. Bag-worms (Psychidae) are moth larvae which build protective cases of silk to which twigs, leaves and fragments of vegetable matter are attached. They do not make a new case at each moult, but enlarge the original one as they grow. They anchor it to a leaf or to the bark of a tree, and close up the opening when they moult. The female is flightless and remains inside the larval case after pupation. The probable function of the case is, again, that it makes the insect distasteful and also disguises it. Pupae of the acacia bag-worm of the Sahel savanna inhabit very tough silk cases which are not only difficult to tear open, but also look much like the seed pods of the trees on which they hang.

Larvae of one of the predatory bugs *(Acanthaspis)* feed on ants and other insects. In addition to stones and pieces of bark, they stick to their backs the empty cuticles of their prey. One of the crabs *(Dromia)* carries a sponge over its carapace. This not only disguises it, but makes it unattractive to otherwise dangerous predators. Mutual associations between different kinds of living organism are the subject of a later chapter, but it can be seen here how they merge into completely different types of defensive mechanism.

Even poisonous plant materials that are merely carried about externally may serve for protection. Larvae of lace-wings sometimes carry on their backs loose packets of plant débris, which may include the sucked-out remains of their insect prey. Although occasionally such packets may serve as camouflage, more often their primary function is one of defence. Ants and other insects that attempt to bite the larva find the packet moved, like a shield, in their direction, and may end up only with distasteful plant material in their jaws. A mature larva can withstand attack from over a dozen ants before its supply of plant débris is depleted. This form of defence is also effective against predatory bugs whose probosces may not be long enough to reach the body of the lace-wing larva even after it has been pushed through the shield. Field observations suggest that predation on the weevils of New Guinea, which have a dense covering of vegetation growing in depressions on their backs, is relatively uncommon, even when they are released in places where they are conspicuous. Camouflage appears therefore not to be the principal function of the covering: it is even more important because it protects the weevils which wear it, and makes them unattractive as food for other animals.

The abdomen of most spiders is sac-like, without visible structure and, though covered by a hardened cuticle, is usually much softer than the front part of the body or 'cephalothorax'. In some primitive spiders, the surface of the abdomen is armed with a series of hard transverse plates, each set with erect black spines. Orb-web spiders sometimes have the dorsal surface of the abdomen covered partly or entirely by a hard plate while, in other species, it is armed with curious spines and processes – some of them of great length. The reason for the development of these abdominal structures is no more apparent than is the significance of similar spines on the cephalothorax. Perhaps, because of the many sharp projections, these spines may discourage birds from attack. Additionally, they may serve to break up the outline and thus help to camouflage the spiders. In some sedentary species, the abdomen is drawn out into a long tail, which gives the animal an oddly worm-like appearance.

Spines are more common in water than on land. The long spines of the larvae of crabs and other crustaceans not only help to keep them afloat or, rather, slow their rate of sinking when they are not actively swimming, but also jam in the gullets of predatory fish and are therefore defensive. Long spines may also be a deterrent against predators other than fishes. One of the rotifers or 'wheel-animalcules' – microscopic inhabitants of freshwater pools – may have long or short spines, or even be spineless. Experiments have shown that individuals with long spines are seldom eaten when contacted head-on by predacious rotifers of another species, while individuals without spines are eaten readily. Furthermore, the time taken to catch and swallow a spiny rotifer is longer than that required to eat a spineless individual, or one with short spines which are far less effective than long spines. When other prey is available, both long-spined and short-spined forms are less heavily preyed upon than are the spineless ones. Spines are only developed in the presence of predatory rotifers which produces a secretion that induces spine formation in the developing eggs of the prey species. When predators are absent, spines are never formed. Although no evidence has been found that long spines impede swimming or feeding, there must be some advantage in being spineless, because otherwise the spined condition would have spread among all individuals and not develop only in the presence of enemies.

Swimmers in tropical waters soon learn to avoid sea-urchins,

Barbs and Spines

whose sharp, poisonous spines are easily broken off and remain embedded in the flesh, causing considerable discomfort and pain. The long spines of sea-urchins are not effective against all predators, however, and sea-urchins have other defences as well. These include numerous small, jaw-like organs, known as 'pedicellariae', which cover the skeletal armour. The pedicellariae snap on to any small animals which succeed in getting between the spines, and kill the larvae of encrusting animals that might otherwise become established on the armour of the sea-urchin. Some pedicellariae contain poison, and bite the tube-feet of attacking starfishes. People whose fingers have been accidentally bitten by the pedicellariae of tropical sea-urchins have reported instant severe pain, giddiness, difficulty in breathing, paralysis of the lips, tongue and eyelids, and relaxation of the limb muscles. The symptoms can last for as long as six hours and, on rare occasions, have caused death.

The entire upper surface of the crown-of-thorns starfish, mentioned in the last chapter, is covered by inch-long spines coated with toxic mucus and enveloped by an integumentary sheath which secretes a venom that also causes extremely painful wounds, swelling, numbness and paralysis. Despite their large size, these animals are relatively inconspicuous, their colour blending closely with that of the surrounding corals upon which they feed. Most starfishes have solid spines which, in tropical species, may be up to 25 cm or more in length and can cause painful wounds but are not poisonous. A few, including the crown-of-thorns, have hollow, sharp spines which penetrate deeply into the flesh of an enemy. Because of their extreme brittleness they break off readily in the wound. Barbs on the tip cause the broken portions, along with their bags of poison, to work their way deeper and deeper into the body with every movement.

The only predator definitely known to feed on these formidable animals is a large, carnivorous mollusc known as the 'triton-shell', which grows to a length of 40–45 cm. An adult triton takes between 12 and 24 hours to consume a full-grown starfish, tearing its prey open with the rasping teeth of its radula, and ingesting it piecemeal. There has also been a report that a small shrimp *(Hymenocera)* also preys on the crown-of-thorns. This shrimp and its mate dance about on the starfish until the latter retracts its tubefeet, thus releasing the grip of its arms on the coral it is eating. The shrimps then raise one of the unattached arms and feed on the flesh of the starfish through the grooves on its undersurface.

Most marine worms are liberally supplied with bristles or 'chaetae' made of chitin – the same material that forms the basis of the arthropodan integument. In some, these bristles serve merely to assist in locomotion and burrowing, and project some distance from the body; but other worms, including the oval, flattened *Eurythöe* and *Hermodice*, are able to retract or extend their chaetae to a quite remarkable degree. When the living worm is at rest, the chaetae appear to be quite short and are barely in evidence but, when it is irritated, the chaetae are rapidly extended and the worm appears as a mass of bristles. The extent to which these bristles afford protection from natural predators is unknown, but they can inflict painful injuries on people who handle the worms carelessly. The severity of symptoms reported in some clinical accounts lends credence to the belief that the chaetae may be venomous. They are hollow and, at times, seem to be filled with fluid, but so far no venom glands have been described in bristle-worms.

Perch, pike and other predatory fishes often find difficulty in swallowing sticklebacks and other spiny prey. The spines of a stickleback have a locking mechanism which keeps them in a fixed position when they have been erected. The spines jam in the mouth or throat of the predator, which has to spit out its prey and grab it again. Pike often reject sticklebacks altogether, and learn not to attack them in future. When they are offered minnows, three-spined and ten-spined sticklebacks, the minnows are eaten first, while three-spined sticklebacks survive the longest. The spines of the ten-spined stickleback are somewhat less efficient, and this accounts for the more timid behaviour of this species when compared with the three-spined stickleback, as well as for its cryptic coloration and habit of living in dense vegetation. When a ten-spined stickleback is pursued by a predator, it always swims directly away, presenting its tail to the enemy. In this way, the predator is forced to seize its prey tail first, so that the spines jam in its mouth. If the stickleback were swallowed head first, its spines would be less effective.

Predatory cartilaginous fishes, such as sharks and dogfish, do

Barbs and Spines

not possess very heavy external armament; but the skin is thick and covered with a scattering of tooth-like denticles which makes it very rough. These denticles are much enlarged in the jaws to form the teeth of the fish. It has often been supposed that the denticle is the primitive type of fish scale, from which others have been derived, but it is now believed that the earliest covering was a continuous layer of bony armour which later broke into large scales from which denticles were ultimately derived. Sharks are also protected by camouflage, afforded by the colour and pattern of the skin pigmentation.

Some fish have poison glands attached to their spines. One of these, the scorpion-fish *(Scorpaena)*, normally lies concealed on the sea-bed. When disturbed, it expands the pectoral fins, whose ventral surface is coloured a conspicuous yellow and black with blue spots. This is probably a warning display because, if further molested, the scorpion-fish raises its dorsal spines and will even use them aggressively to ram an attacker. Such defence may be effective against predatory fishes, but it does not prevent the scorpion-fish from being devoured by an octopus or squid, so the poisonous spines are not always a deterrent. Indeed, when hunted by an octopus, the only hope of escape for a scorpion-fish lies in flight.

Few detailed studies have been made of the use of poisonous spines against natural predators, but several fish species, in addition to the scorpion-fish, are known to have a poison that is toxic to man. These include sting-rays, chimaeras, toadfishes, catfishes, star-gazers, weevers, gurnards and stonefish. Doubtless it is to some of them that the 91st Psalm refers in the words: 'They shall bear thee up in their hands, lest thou dash thy foot against a stone. Thou shalt tread upon the lion and the adder: the young lion and the dragon shalt thou trample under feet.' If taken literally, it would seem curious that the young lion should be associated with the dragon, and the lion with the adder; and also rather strange that a lion should be trodden underfoot. If, however, the verses are assumed to refer to fishes, the whole passage forms a metaphorical picture: for the lion-fish is another name for the scorpion-fish, the adder is the adder-weever or viper-weever, while the dragon stands for the greater weever *(Trachinus)*, also known as the dragon-fish. All these fishes are inhabitants of the Mediterranean Sea, all lurk under stones, and all are mentioned in earliest writings. For instance, the physician and naturalist Apollodorus referred to them about 300 BC in his book *Stinging and Biting Animals*, which became a work of reference for subsequent writers on this subject. The names dragon-fish and lion-fish are also applied to tropical species *(Pterois)* found on coral reefs in the Indian Ocean and in the south-western Pacific as far east as Samoa, but it is unlikely that the psalmist was referring to these. The father-lasher *(Cottus)* of the Atlantic is also sometimes known as the sea-scorpion or lion-fish.

A surprising variety of marine fishes are protected by poisonous spines. Even the common mackerel has a small spur between the anal fin and the vent, a puncture from which can cause a festering wound; while a related species from Maderia has a double-grooved dagger in the same region of the body which induces symptoms so severe that they have sometimes necessitated amputation of the affected part. The poison of the weevers causes very acute pain, followed by severe inflammation; that of the spiny dogfish and sting-ray has similar effects. The agony may be so great that fishermen who have been stung have tried to throw themselves overboard, and there is at least one record of a man cutting off his own finger to obtain relief. People with heart compaints have reportedly been known to die within a few minutes of being stung by the lesser weever, and the suggestion has been made that the powerful poison of this fish has been responsible for many of the cases of mysterious drownings that have occured in British waters. Immediate excruciating pain, followed by shock, oedema, paralysis and loss of sensation is characteristic of the sting of most venomous fishes. General effects often involve respiration while, in extreme cases, the victim falls into a deep coma and never recovers consciousness. Abscess formation, necrosis and gangrene frequently delay convalescence, while dyspnoea and general weakness may persist for several months after a sting.

Stonefishes *(Synancega)*, especially notorious for their dreadful stings, are a particular menace to bathers and people wading in tropical waters. Sluggish, protected by a thick covering of slime on their irregular, warty skin, they become coated with coral débris, mud and algae, and are consequently extremely difficult to discern. Their wonderful camouflage is enhanced by their habit of burying themselves in sand just as weevers do. It is surprising to find effective concealment associated with such powerful defences. Most poisonous animals advertise their presence with conspicuous warning coloration so that they are avoided. Of course, the poison spines of fishes have not been evolved in relation to man, so it may be that camouflage is effective against certain predators that hunt by sight while the poison spines are only used when the fish has been detected or else to deter predators that use the senses of smell or touch. Despite their concealment and poison, stonefishes are by no means immune to predatory enemies. They are sometimes eaten by bottom-feeding sharks and rays, apparently without causing ill-effects, and young ones quite often fall victim to large conches.

Fishes are not the only vertebrates to bear defensive poison spines. Both the duck-billed platypus and the echidna – sole surviving representatives of the egg-laying monotreme mammals – are equipped with poisonous spines on their back legs. The venom apparatus is found only in the males. It consists of a movable horny spur of the inner side of each hind limb, near the heel. The spur projects about 15 mm and is tapered, conical and slightly curved. Its base is enclosed in a fleshy sheath and, although it normally lies against the leg, it can be erected into a right-angled position. The poison gland of the platypus becomes active during the breeding season: the glands of echidnas, on the other hand, are never very active.

None of the poisons mentioned in this chapter have any relation to feeding. Like the repugnatorial secretions to be discussed in the following chapter, they are concerned only with defence: they must have been evolved quite separately from the stings and venoms that are associated with the capture of prey, but which may also have a secondary defensive function. Their origin doubtless lies in defensive barbs and spines whose efficiency has been enhanced by the evolution of associated poison glands. In this way, enemies can be repelled and driven away by means of smaller and less cumbersome, but far more devastating weapons. Poison is a remarkably effective, as well as an economical weapon of defence.

6 Chemical Defences

Many animals rely for defence upon special glands, the chemical secretions of which are so thoroughly distasteful, smelly and generally unpleasant, that almost no potential predator will have anything to do with their possessors. Such 'repugnatorial' glands, as they are called, have aroused considerable interest among biologists; and a survey of the literature reveals an abundance of information about the secretory discharges produced by millipedes, insects, arachnids, fishes, amphibians, reptiles and mammals when roughly handled or otherwise disturbed. A well-known example is afforded by North American darkling beetles *(Eleodes)* which stand on their heads when annoyed and spray their enemy from glands at the tip of the abdomen with a smelly secretion that irritates the skin. This substance contains benzoquinones: it is the chief line of defence of these insects and has been found to repel a diversity of predators most effectively.

The first toxic substance to be isolated from an insect was formic acid. Produced by ants, this was chemically analysed in the seventeenth century. Anyone who has disturbed a nest of wood-ants will have noticed the acrid spray ejected by these insects. The second toxin was cantharidin, secreted by blister-beetles, and first isolated in 1810: but most of our knowledge of arthropod poisons has been obtained during the last twenty or thirty years.

Defensive secretions may take their effect by other than purely chemical means. The materials discharged are sometimes sticky or slimy and, as a result, entangle or mechanically restrain an attacking predator. An odourless fluid that hardens to a rubbery consistency on exposure to air is ejected by members of the Onychophora – a group of carnivorous animals, related to the arthropods, and which look rather like caterpillars. This provides

Chemical Defences

an effective defence against attacking centipedes, spiders and insects. The discharge is aimed with precision, and even single ants can be hit and stuck down, for several minutes at least, while some never escape. Comparable defensive mechanisms are found in certain centipedes which extrude sticky threads from their posterior legs when attacked by ants or spiders, and in spitting spiders *(Scytodes)* which squirt poisonous gum from their rapidly vibrating jaws whilst on the move, after the manner of a cruiser tank spraying infantry with machine-gun fire: the prey is stuck firmly to the ground while the spider advances and eats it at leisure. The gum is used defensively as well as in offence.

The slime of earthworms and slugs renders such creatures relatively free from attack by carabid beetles, ants and centipedes; while certain cockroaches secrete a layer of slime at the rear end of the abdomen which serves them in a similar manner. It is well known that droplets of fluid ooze from the tips of two peg-like 'cornicles' on the abdomen of aphids when these animals are disturbed, and there seems little doubt that this mechanism must be defensive also. The fluid apparently consists of droplets of wax in a liquid state which changes to a solid crystalline phase on con-

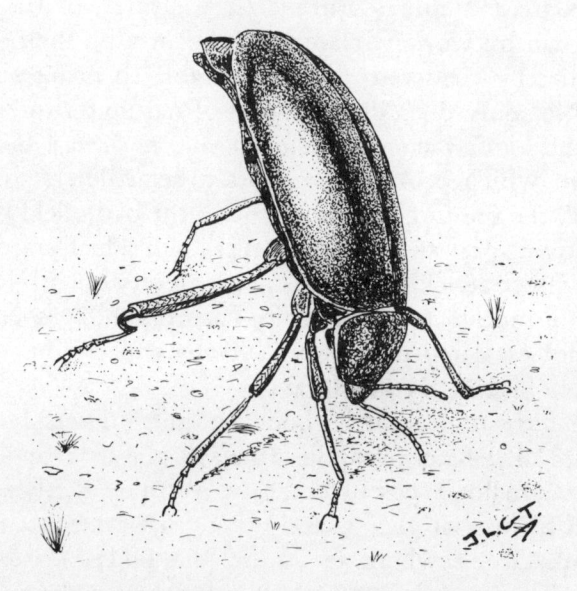

16. The North American darkling beetle *(Eleodes)* stands on its head when disturbed, and sprays a defensive secretion at the attacker from glands in the abdomen. (After W. Wickler.)

tact with extraneous surfaces, although the melting point of the wax is actually higher than that of normal summer temperatures. Aphids can sometimes be found with the shrivelled bodies of parasitic hymenopterans attached to them by plaques of cornicle wax. In addition to the cornicles, aphids possess other weapons of defence against their natural enemies – lace-wing larvae, ladybird beetles and so on. Special glands on the body surface secrete a dense coating of flocculent wax filaments, the integument of the body may be heavily sclerotized, and the hind legs modified for jumping. Such defences tend to be poorly developed, or even absent, in species that are tended and protected by ants.

Sticky fluids are also used in defence by certain termite soldiers known as 'nasutes', whose weapon is a spray secreted by the frontal gland and ejected from the pointed nozzle or 'rostrum' of the head. The spray is a viscous entangling agent which rapidly hinders the mobility of ants. At the same time, it also acts as an irritant, inducing scratching and preening which, in turn, cause the sticky contaminant to be further spread over the surface of the enemy. Worker termites have no special weapons, but they can bite effectively and play their part in the defence of the colony. Ants are sometimes crushed by the bites; or they may be slowed down by workers clamped to them with their mandibles and are thereby rendered more vulnerable to being sprayed by soldiers. Not only does the secretion of nasute termites entangle enemies and block their sense organs, but its scent acts as a powerful alarm which promptly recruits other soldiers to the fray. On arrival, the reinforcements surround the battlefield but do not add their own spray to the target unless individually prompted to do so by direct attack from the enemy. Consequently, only ants that are still mobile and potentially harmful are sprayed, and the battle is not escalated uneconomically by superfluous discharges themselves eliciting further sprays.

Among terrestrial animals, none are more diversely endowed with chemical defence weapons than are the arthropods, which possess two major types of defensive substances: those that are elaborated by special glands, and others that are contained in the blood, gut, or elsewhere in or on the body. Caterpillars of swallowtail and parnassian butterflies possess a defensive gland situated dorsally behind the head. This consists of a two-pronged invagination of the neck membrane, which can be abruptly everted when the insect is disturbed. The extruded 'horns' glisten with an intensely odourous secretion. The larva arches its body and wipes them directly against its enemy – with such accuracy that even small predators, such as ants, are almost always contacted and repelled. Similar eversible glands are to be found in beetles and other insects.

Although millipedes are, on occasion, eaten by a variety of animals, including toads and birds, to the majority of would-be predators they are unpalatable because of their tough integuments and the irritant exudates secreted by their repugnatorial glands. In most cases, the secretion is exuded fairly slowly but, in some of the larger tropical forms, it can be discharged to a considerable distance in the form of a fine jet or spray. The fluid has a strong caustic action, and causes blackening on contact with the skin. Later, the affected part peels, leaving a wound which heals only very slowly. The secretion is dangerous to the eyes and has been responsible for numerous cases of blindness among chickens, in the West Indies and elsewhere, that have misguidedly pecked at passing millipedes. The chemical compounds chiefly responsible are iodine, quinine, hydrocyanic acid and small amounts of chlorine.

Centipedes are not protected solely by their venomous bites and sticky secretions. Several of the many-legged Geophilomorpha excrete a phosphorescent liquid, when disturbed, which serves as a protection against attack by ants and other enemies. In Southern California, Arizona and New Mexico, the large, greenish scolopendras are feared, not only on account of their poisonous bites but also because they produce a reddish streak where they have crawled over the body. Like many other subtropical and tropical scolopendras, they make tiny incisions with the claws of their 22 pairs of legs. In themselves, these are trifling but, when alarmed, the centipedes drop into each incision some kind of poison that causes intense irritation, so that the affected part becomes inflamed and the two rows of punctures show white against the skin. No doubt such phosphorescent and poisonous fluids, which are secreted near the bases of the legs, are protective and render centipedes distasteful or even dangerous to their enemies.

The whip-scorpions of Mexico and the southern United States have glands that open on a veritable gun emplacement – a revolving knob at the rear of the body. The name 'vinegaroon' was originally bestowed on them by settlers from the French West Indies, and arose from the vinegar-like acid secretion they exude when disturbed. A blacksmith, in 1877, is reported to have inadvertently crushed one to his upper left breast: the blisters which resulted from the repugnatorial fluid of the whip-scorpions extended over the whole of his chest!

Aiming of repugnatorial secretions may be achieved in a variety of ways. Earwigs flex their abdomens which bear glands at the base, while some ground-beetles (Carabidae) spray a secretion containing formic acid from two glands that open on the rear of the abdomen beside the anus. These beetles spray from one gland or the other, depending on which side of the body has been stimulated, and control the direction of the discharge by twisting the tip of the abdomen. The large two-striped walking-stick *(Anisomorpha)* of south-eastern North America is exceptional in that it sometimes discharges its spray at approaching birds before these have actually pecked it; and it is clear that the sense of sight must be involved in recognition of the enemy.

The behaviour of the giant South American cockroach *(Blaberus)* alters according to its stage of development. Nymphs tend to bury themselves or hide in crevices wherever possible, whereas adults do not do this to any great extent. Burrowing by both adults and nymphs takes place significantly more often in light than in darkness but, whereas nymphs avoid the light at all times, adults do so only in the resting phase of their activity cycle. These differences in behaviour result in the nymphs being far more secretive than adults. So it is really not surprising to find that the adults have a horrible smell, while the nymphs are odourless and the spines on their legs are not sharp like those of the adults. Having no defence, apart from their speed of running – for only adult insects are winged – the cockroach nymphs must survive entirely by their anchoritic habits.

The milky secretions of the thoracic glands of water-beetles are toxic. Fish are partly narcotized and caused to vomit as a result of eating them, while amphibians and small mammals also find these beetles emetic. Whirligig-beetles also secrete a viscous fluid which causes fishes and newts to reject them as food. Various pentatomid bugs have defensive glands in the thorax of the adult and the abdomen of the larva which can be discharged to the left or right, according to the direction of an attack.

The contents of defensive glands are normally expelled either by contraction of appropriate muscles that surround the reservoirs of the glands, or by compression of the glands themselves through a rise in blood pressure effected by some sort of contraction of the whole body. Pacific beetle-cockroaches *(Diploptera)* rely on the pressure of air in the respiratory system to discharge their repugnatorial glands, while certain grasshoppers emit a poisonous froth that consists of a mixture of glandular secretions with air from the tracheal system.

Predatory bugs which normally inject venom from their salivary glands into insect prey, can also spray it for a considerable distance in response to attacks by larger predators: while assassin-bugs *(Platymeris)* spit a stream of saliva from the rostrum when disturbed. This saliva contains a poison which quickly kills their prey when injected, and has an action resembling that of snake venom. Since arthropods are unaffected by topical application of the saliva, however, spitting must be a defence against vertebrate predators. If the saliva comes in contact with the mucous membranes of the eyes or nose, it causes intense local pain, swelling, vasodilation and bronchial disturbance.

Normally, arthropod defensive glands consist of membranous, sac-like invaginations of the body wall containing secretory glands in the wall of the sac itself, or as distinct clusters of cells connected to the sac by way of a duct. Certain species, however, have exceptional glands in which the stored products are not the final constituents of the secretion, but merely its chemical precursors. These glands are appropriately named 'reactor' glands because the precursors are not mixed until the moment of discharge. They are found in certain millipedes and beetles. In the former, benzaldehyde and hydrogen are dissociated from mandelonitrile by means of an enzyme stored in a separate compartment of the gland, and are liberated as vapours that enshroud the animal when it is under attack. The reactor glands of beetles generate benzoquinones by mixing hydroquinones and hydrogen peroxide from the glandular reservoir with appropriate enzymes generated in an adjacent compartment.

South American bombardier-beetles, on being seized, immediately discharge a defensive secretion that is so hot that only a few specimens can be captured at a time with the naked hand. The spray is secreted by a pair of reactor glands that open at the tip of the abdomen. The inner reservoirs contain an aqueous solution of hydroquinones and hydrogen peroxide, the outer compartments a mixture of catalases and peroxidases which decompose the hydrogen peroxide and oxidize the hydroquinones. Oxygen and heat are produced as a result of these reactions, and the repugnatorial spray reaches a temperature of 100°C at the moment of discharge!

Other ground-beetles secrete a variety of defence substances. Unique among these is one produced by female *Pterostichus* and directed towards the males of the species. When a female is receptive, she will usually mate within 5 minutes of being found by a male but, if unreceptive, she will run from him. When pursued, however, she discharges a blast of liquid from the tip of her abdomen towards her ardent suitor. On receiving the discharge, the male immediately stops running. Within 10 seconds, his movements become so unco-ordinated that he may roll on his back and be unable to regain his footing; his legs become stiff, and movement ceases. He remains in a death-like coma for up to 3 hours but eventually recovers, apparently none the worse for the traumatic experience. The female is immune to her own secretions. It is believed that this repugnatorial secretion is used to prevent males from eating the eggs at the time of laying, and may also be effective against other arthropodan enemies.

Most repugnatorial secretions are effective against invertebrate and vertebrate enemies alike. The ability of some insects to aim their defensive sprays enhances the efficacy of the weapon, especially when it is used against smaller predators that might otherwise easily be missed. The secretion is often ejected so promptly that it prevents the predator from causing a disabling injury to its prey. Although vertebrate enemies do sometimes inflict considerable harm, they often learn to avoid similar prey after a few encounters, especially when the prey shows distinctive coloration. At the same time, no defence is perfect, and American grasshopper-mice learn to subdue quinone-secreting darkling beetles by holding them head up and forcing the tip of the beetle's abdomen downward so that the repugnatorial secretion is discharged harmlessly into the soil.

The effects of repellent exudations are usually of short duration: the predator releases its hold and flees instantly before

Chemical Defences

cleansing itself of the repugnatorial fluid. Only in exceptional cases, such as the secretions of some nasute termites, are these compounds known to be fatal: most of them merely cause temporary irritation. Sometimes, however, they may be viscous and sticky, as well as distasteful and smelly, and then they can have a direct mechanical effect – especially on small predatory arthropods.

In addition to the secretions of repugnatorial glands, many arthropods possess defensive chemicals in their blood and may have active control over its release when they are attacked. Reflex bleeding from the points of the legs is found in a number of beetles. When a single leg is pinched, as an ant might do with its mandibles, a droplet of blood is released from that particular leg, while a localized stimulus applied to the body of the beetle elicits a response only from the nearest leg. The blood provides effective protection against ants which have, no doubt, been a major agent in the evolution of this defensive mechanism.

Reflex bleeding from blister-beetles is an effective deterrent to predacious arthropods and vertebrates alike because the blood contains cantharidin, an extremely powerful vesicant. At one time, cantharidin, prepared from the wing-cases of the 'Spanish fly' *(Lytta)*, was widely used as an external irritant and, until its dangerous properties had been appreciated, as an aphrodisiac. Its chief use today is in hair-restorers, but during the American Civil War, it was highly esteemed for its use in love potions. When their usual supplies were cut off from the Confederates by the blockade, a substitute was obtained from a species of potato-beetle belonging to the same family (Meloidae).

Insects with protective chemicals in their blood sometimes have separate defensive glands as well. These contain the same toxic principle as that found in the blood. Some insects defaecate in defence, while grasshoppers regurgitate a fluid from their crops that is toxic to mammals: it is an irritant to the eyes and induces vomiting when swallowed. The regurgitated fluid has been shown experimentally to disperse attacking ants, but tests with captive jays have shown that these birds have a rather elegant way of circumventing the defences of grasshoppers. They peck the head first, then pull it away, together with the crop and its contents. Subsequently, they eat the body of the insect which is perfectly edible.

Tooth & Claw

There are two principal ways in which an animal comes to possess a defensive chemical: it can synthesize the material itself, or it may sequester it from some external source, such as the plant material on which it feeds. Thus, the cinnabar moth contains senecio alkaloids, presumably obtained from plants of the genus *Senecio* upon which the larva feeds. Grasshoppers of the genus *Poekilocerus* are extremely poisonous to birds after eating their normal food – milkweed plants of the family Asclepiadaceae, which contain cardenolide toxins. When they have been fed on cabbage or dandelions, however, these insects can be ingested with impunity. Similarly, when monarch butterflies, which normally feed on Asclepiadaceae, are reared on cabbage, the larvae, pupae and adults, are equally palatable to bluejays, to which they are normally toxic and emetic.

Secondary plant substances need not be incorporated systematically in order to provide protection. It may be sufficient for them to be taken into the gut and subsequently regurgitated defensively. The degree to which fluids regurgitated by grasshoppers repel attacking ants depends very much upon the food plants of the insects. Certain aromatic plants, however, themselves distasteful to most insects, appear to be inactivated enterically by grasshoppers; and the intriguing suggestion has therefore been made that these insects may select their food not only in relation to their ability to cope with its toxins, but also according to their ability *not* to detoxify ingested poisons which may be preserved for purposes of chemical defence!

Little is known of the extent to which defensive substances are transmitted from one substance to another in a food chain, but there is no doubt that such transmission must occur. After eating the meat of frogs captured locally, French soldiers in Algeria during the last century sometimes developed urogenital symptoms including 'érections douloureuses et prolongées', such as were known to doctors at the time to be characteristic of cantharadin poisoning. The frogs were shown to come from areas where blister-beetles were abundant, and examination of their gut contents showed them to have been feeding on the beetles. The legionnaires were getting more than they bargained for, an aphrodisiac with their epicurean delicacies!

Many species of fish become poisonous through feeding on marine organisms, especially plant plankton, which are not toxic

Chemical Defences

to the fish but produce poisonous substances in their flesh. Carnivorous fish that feed on these herbivorous fish then become poisonous themselves. Poisoning that results from eating such fishes is known as 'ciguatera' poisoning and has been attributed to over 300 kinds of fish, mostly reef species in the Pacific ocean and West Indies. The most poisonous fish in the world is the Japanese puffer-fish, or 'fugu', whose internal organs, skin and muscles, contain a very powerful nerve poison called 'tetrodoxin'. This causes respiratory paralysis, from which most human victims die within 24 hours, and there is no known antidote. Extraordinary as it may seem, the flesh of this fish is highly esteemed and special cooks, who are skilled at removing the poisonous parts of the fish, are employed in the restaurants where it is served. Despite this, however, the puffer-fish is the main cause of food poisoning in Japan and there is a saying in that country: 'Great is the temptation to eat fugu, but greater is the dread of losing life.'

Frogs and toads often have defensive glandular secretions, sometimes associated with warning colours. The chemical nature of the toxins is extremely variable, even within the same species, probably depending upon the extent to which they are preyed on. In the case of the Panamanian poison-frog *(Dendrobates)*, for instance, the toxicity of members of local populations is not related to their coloration, since some apparently cryptic races are even more poisonous to mice than are more brightly coloured races. The 'batrachotoxins' produced by these frogs are reputed to be the most powerful known naturally occurring nerve poisons, while the skin of salamanders contains another deadly alkaloid neurotoxin known as 'samandarin'.

The common European grass-snake struggles violently when first caught, and then discharges from its vent an evil-smelling liquid secreted by the anal glands. The secretion is formed by breakdown of the cells which form the epithelial lining. It is thick, somewhat oily in consistency, and varies in colour and smell. Its exact composition is unknown, but the disgusting odour, so difficult to wash away, is well known to all who have handled these snakes. All snakes possess anal glands, but the fluids of most species are odourless to man, and not every snake voids its secretion when trapped, as the grass-snake does. Whether the liquid serves as a defence mechanism against the

Tooth & Claw

snake's natural enemies has not yet been established, but it seems probable that this must be the case. The odour also plays a part in the courtship and mating of snakes.

The slow-worm and other reptiles often defaecate defensively, while tortoises discharge their smelly urine. Tortoise urine not only protects the individual that excretes it, but is also used to protect the eggs. After oviposition, the eggs are buried and the female tortoise urinates on the sand or earth that covers them. This either prevents the smell of the eggs from being detected or else renders them unattractive to the dogs, hyenas, coyotes and other animals that would otherwise dig them up.

In birds, odour is not very noticeable, although some species give off a characteristic musky smell. The hermit-ibis *(Geronticus)* is said to reek like putrid carrion, and its flesh tastes as bad as it smells. Brightly coloured birds are often unpalatable, while cryptic and vulnerable species enjoy a wide reputation for their flavour. In a series of experiments carried out in Egypt by Hugh Cott, in which hornets were offered choices of the freshly skinned carcasses of various species, the two which attracted most insects were the wryneck and crested lark: these were also the two most cryptic in appearance. Those of lowest edibility, as assessed by the numbers of hornets taking meat from them, were chats, pied kingfishers, masked shrike and hoopoes – all with very conspicuous coloration. Similar results have been obtained using cats and foxes as well as human tasting panels to assess edibility. Clearly, distastefulness in birds is more than skin deep. Again, experiments with egg-eating mammals such as cats, ferrets, mongooses, rats and hedgehogs, have disclosed that the most ill-favoured eggs are commonly either immaculate white or white with reddish spots, while eggs protected by cryptic coloration, laid in colonies or in sites difficult of access to predators, are usually the most palatable.

When threatened by danger, birds of many species react by vomiting or defaecation. To what extent the offensive odours emitted in this way have more than an incidentally repelling role is difficult to estimate, although a human being would certainly take steps to avoid a squirt of stomach contents from an albatross, or of faeces from an eider duck! Many kinds of duck, when flushed from their nests, protect their eggs by defaecating on them, but whether the foul smell gives protection against egg-

Chemical Defences

eating mammals is doubtful. It has even been suggested that the scent may inadvertently help predators in finding nests.

Discharge of foetid secretions for defence is also practised by mammals of several different families which use the products of the perineal glands for this purpose. Many examples can be found among the civets, mongooses and weasels, but it is among the skunks that this method of defence reaches its greatest development. The anal glands of these otherwise attractive creatures are very large indeed, and the substance they secrete can be shot out, with astonishing precision, to a distance of 4 metres! Nocturnal in their habits, skunks roam about at dusk in search of insects and other small animals on which they feed. They move in a nonchalant, leisurely manner, as though well aware of their ability to defend themselves, holding up their white tails as a warning against approaching within range of their nauseous emissions. Before discharging their repulsive liquids, skunks first assume threatening attitudes which tend to vary both with the species of skunk and with the enemy to which it is responding.

17. Positions adopted by skunks *(Mephitis* and *Spilogale)* before ejection of their malodorous secretions. (After F. Bourlière.)

Striped skunks *(Mephitis)* stamp on the ground with their hind feet while other kinds, such as the spotted skunk *(Spilogale)*, walk for a few seconds at a time on their front legs with the hindquarters elevated. The possession of stink glands and the ability to deliver a gas attack is advertised by conspicuous black-and-white patterns of coat coloration. By deterring an aggressor in this way, the skunk can economize in the use of its secretion.

The South African zorilla *(Ictonyx)* is yet another member of the weasel family. Its black body bears on the back a set of broad, conspicuous white longitudinal stripes, like the stripes on the face of a badger. If interfered with, the zorilla enhances its already conspicuous appearance by erecting the hairs of its body, brandishing its bushy white tail, at the same time uttering shrill squeaks of anger and ejecting an intolerable acrid secretion from the anus. Badgers and ratels, which also possess stink glands, are slow and leisurely in their movements, fearless in manner, and conspicuous. The teledu *(Mydaus)* of Malaya and Indonesia is related to the zorilla, and has similar warning coloration. It is nocturnal, slow, and largely insectivorous. Again, it possesses stink glands which exude a foetid liquid whose smell is so horrible that it has been known to cause people to faint! In South Africa the grison, a relative of the polecat, has somewhat similar coloration and, if disturbed, emits a disgusting stink, fights savagely, and is extremely tenacious of life.

Many mammals possess scent glands with which they normally mark their territories. These glands are sometimes also discharged in stress situations. The peccary, for instance, has one on its back, in front of the tail. When the animal is excited, the hairs of the back and neck bristle and the dorsal gland emits a musky secretion which can be smelt by man from a distance of several hundred metres. It would seem that there are many intermediate stages between the emotional discharge of cutaneous glands in response to fear, and their utilization for defence against an aggressor; but one can see how the first might have evolved into the second. Even a faintly unpleasant smell might be enough to deter a predator that was only slightly hungry, and the stronger the odour the more effective it would be. Consequently, natural selection would tend to favour those individuals that produced the most unpleasant secretions, and they would survive to pass this hereditary characteristic to their offspring.

Chemical Defences

Unique among animals is the ability of certain fishes to deliver electric shocks. About 250 species, often unrelated, are known to possess electric organs. Among these, the most interesting are perhaps the torpedoes, star-gazers, skates and rays of temperate and tropical seas; and the fresh-water mormyrids or 'snout-fishes' of Africa, the electric eels *(Electrophorus)* of South America – these are not true eels but are related to the carps – and the electric catfish *(Malapturus)* of rivers and lakes in tropical Africa. The structure and position of the electric organs of these fishes differ greatly. Each organ is, however, made up of a number of disc-like cells or 'electroplates' embedded in jelly and bound together by connective tissue into an elongated tube. One side of each electroplate is supplied by nerve fibres and the jelly-like material is well provided with blood vessels. The plates are arranged in blocks that combine to produce the electric shocks.

All activities in the animal body are accompanied by electrical changes – changes so minute that they can only be detected with the aid of special instruments. In nervous tissue they are used to transmit messages within the body: but they can be intensified and utilized for other purposes. In some fishes they are intensified to such a degree that they can be used to give dangerous electric shocks. The blocks of electroplates are like the various cells of a high-power battery. Individually, their voltages are low, but they can be added to one another to produce a high voltage.

The black torpedo generates an electric discharge of about 60 volts, but other rays, skates and mormyrids do not produce more than 4 volts. During the day the torpedo rests on the sea-bed, partly buried in sand, and stuns other fish that swim too close to it. At night it cruises over the bottom, and stuns fish below it on the sand. A discharge of up to 450 volts has been measured from the electric catfish, while a record 650 volts has been registered from an electric eel weighing 40 kg! This would be sufficient to kill a man on contact, or to stun a horse at a distance of 6 metres. In 1941, two Indians were killed instantly in the State of Amazonas, Brazil, when they slipped while walking along a plank, and fell into a pool containing electric eels. Electric shocks are used both in defence and to capture prey otherwise too large or too swift for the electric fish to tackle. Marine electric fishes do not need such very high voltages to stun or kill, because salt water is a much better conductor of electricity than fresh water.

Tooth & Claw

The significance of weak electric discharges has only been understood in recent years. Fishes like mormyrids, which produce a series of electric pulses, use them to create an electric field around themselves. Should any object break this field, the fish immediately becomes aware of the change. Since most of the snout-fishes live in muddy waters and have poorly developed eyes, the value of an electric field as a warning device signalling the presence of predators, prey or inert objects, is immediately apparent. These fishes can find their way about and detect obstacles in complete darkness. Charles Darwin had been at a loss to explain the evolution of powerful electric organs, because he could not imagine any biological advantage (which could be progressively strengthened through natural selection) accruing to electric organs too weak to influence predators or prey. When it is appreciated that a weak organ may act like a radar warning system, however, part of the problem is solved – even though the possible functional advantages of electric discharges of intermediate strengths have yet to be explained. Sharks, rays and catfish *(Ictalurus)* are able to locate living food by means of electric sense organs which detect muscle action potentials – the feeble electric discharges which accompany muscular action. They are even able to discover fishes resting passively on the sea-bed or hidden under sand. Some mormyrids use electric discharges for intra-specific signalling but not apparently for locating or stunning prey.

In addition to its powerful discharge, the electric eel also emits continuous weak pulses which, like those of mormyrids, act as an electrical warning and navigational device. The organ of the electric rays may serve a similar purpose, but the function of the small electric organs of other rays remains a mystery, because they do not produce any continuous discharge. The fish can only be stimulated to produce its electrical current by vigorous prodding, and the discharge is far too weak to be an effective deterrent to predators.

During the last century, the natives of Calabozo, Venezuela, used to catch thousands of electric eels by a method known as 'embarbascar con caballas ' (intoxication by means of horses). These unfortunate animals were driven by shouts and blows into the waters of the rivers where, according to Charles Holder, writing in the last century, they would dash about as if aware of

Chemical Defences

their danger. 'Great eel-like yellow bodies appear, their backs flashing in the sun, darting about, hurling themselves against the terrified beasts, which with staring eyes and trembling frames are completely paralysed by the electric discharges. Some are killed as if by lightning, and fall among the writhing mass; others endeavour to break through the howling throng of natives upon the banks, but are beaten back to terrible death or torture. The eels seem to be aware of the most vulnerable points to attack, as they strike the poor brutes near the heart, discharging the whole length of their battery. The terrible struggles last from twenty to thirty minutes, and then those horses that have survived the ordeal seem to grow careless of the attacks. The fishes have exhausted their electric supply for the time; and now the natives step to the fore. The eels, finding their power on the wane, seek the bottom of the lake; the natives, mounting the horses, rush wildly about among the fleeing animals, striking them with their long spears, and dragging them ashore, or anon rolling from their horses, paralysed by unexpected shocks that dart up the wet lines. Great numbers of eels are captured, and it is always found that, though they soon exhaust their force, if an attack is intended the next day the same precautions are necessary, their recovery of vital force being extremely rapid.'

Embryological studies have shown that the electric plates of torpedoes have been developed from some of the branchial muscles which have been relieved from their original duty of moving the gills, while the organs of skates, electric eels, catfish and mormyrids owe their origin to transformations of some of the lateral muscles of the body. In star-gazers, the electric organs have been developed from portions of the eye muscles, each of the plates representing a single muscle fibre.

Unlike repugnatorial secretions, electric shocks can be used both for defence and to capture prey. In this respect they approximate more to venoms delivered by stings and teeth, which may possess a dual function – though whether their initial use was in defence or offence is often far from clear. As in the paradox of defensive structures and the appropriate behaviour which accompanies them, both must have evolved simultaneously. Much depends upon the outlook of the observer. We tend to think of chickens reproducing themselves by means of their eggs, but it would be equally valid to regard the chicken merely as a mechanism by which one egg begets another! It is even possible to argue that the egg came *before* the chicken. This is because the eggs of birds and reptiles, like those animals themselves, are well adapted to terrestrial conditions. In particular, the developing embryos excrete nitrogenous wastes in the form of insoluble uric acid. This is not toxic, as the ammonia excreted by the embryos of fish and amphibians would be if it were not removed by the surrounding water. Nor does a deposit of uric acid crystals upset the osmotic balance of the egg, as a concentration of urea would. The urine produced by mammalian embryos, of course, is excreted through the kidneys of the pregnant mother. Uric acid metabolism is retained by birds and reptiles throughout life, and enables them to excrete a very concentrated urine without unnecessary expenditure of water. This is a factor which enables reptiles, especially, to become such successful desert animals. But it could be argued that, in the first instance, the excretion of uric acid was evolved in the egg, enabling it to develop on land and, consequently, that the chicken's egg came before the chicken!

We have strayed somewhat from the subject of electrical discharges from fish, but the argument illustrates the point that all evolutionary adaptations and changes involve 'feed-back' between the evolving organism and its environment. As an animal develops a particular defence mechanism in response to predation, the predators change their modes of detection and attack. In consequence, further modifications are produced in the prey animals as a result of the change in selection. So it goes on – a continual and never-ending series of responses and interactions between animals and every aspect of the environments in which they live. A steady state can never be reached for long, if at all, because some aspect of the environment, even a climatic factor, is sure to change and upset the balance.

7 Venoms

The word 'venom' calls to mind the occult, the sinister, the more diabolical modes of inflicting injury. Venoms and poisons feature predominantly in superstition, folklore and legend. Some of the oldest medical writings, including the Ebers Papyrus, the Smith Surgical Papyrus, and the Hearst Medical Papyrus, describe prescriptions for treating bites and stings which were in vogue about 3,700 years ago. Zoologists, on the other hand, regard venoms in a more mundane light. They are merely defensive compounds that certain animals secrete and later communicate by biting, stinging, or through mere contact with other animals. Some poisonous substances have been evolved for defensive purposes only: others, such as snake venom, no doubt appeared initially as a means of subduing the prey; but they, too, are often associated with conspicuous, warning coloration, and are effective deterrents to potential enemies. With the exception of social Hymenoptera – wasps, bees and ants – venomous animals tend to have solitary habits.

Unlike the secretions of repugnatorial glands, venoms that are injected by teeth or stings are usually ineffectual when merely applied to the skin or taken into the alimentary canal. It is even possible to suck the poison from a rattlesnake bite without suffering ill effects, unless the person administering first aid in this way has a cut on the lips or an abrasion of the mucous membranes of his or her mouth. Only when they are injected through the integument, or come in contact with cut surfaces, do venoms normally induce their characteristic reactions. Consequently, devices are required for injecting them, and laceration or penetration of the skin or integument are necessary for venoms to be manifested. Possibly other glandular secretions might be considered to be venoms if they were accompanied by mechanisms for their introduction into the tissues of an enemy or prey.

In many instances, venoms are injected by the teeth or mandibles. This is so in the case of poisonous snakes, of course, but it is also true of insects and other animals. The mandibles are the principle weapons by which ants attack their prey and defend themselves from the unwelcome attentions of their enemies. In some cases bites and stings supplement one another. Wood-ants not only spray their enemies with formic acid, as mentioned in the last chapter but, by moving the abdomen forward, inject it into wounds made by the mandibles. In battles with ants of other species, or with members of other colonies of their own species, wood-ants pull their opponents' legs and antennae with their mandibles, and spray with acid the stretched membranes exposed between the segments. Sometimes enough formic acid is absorbed to cause paralysis, or even death of the enemy. Some ants bite and sting simultaneously. The bite is necessary to provide a mechanical advantage while the sting is being inserted, but salivary secretions are not introduced into the wound. It has been shown experimentally that the bulldog-ant of Australia is unable to insert its sting if the mandibles have been removed, and the same applies to the American fire-ant.

The hypodermic syringe mechanism of wasps, bees and ants is a most sophisticated device for administering venom. It is employed defensively when the colony is threatened, but has its origin in the needs of nutrition: wasps and some species of ants use their stings for killing prey. In the case of wasps this is chewed up and fed to the larvae.

The ability of Hymenoptera to give painful stings led many early scientists to investigate the venom apparatus. These included R. A. F. Réaumur and J. Swammerdam who, in 1740 and 1752 respectively, described the very complicated sting mechanism of these insects. The sting apparatus of an hymenopteran consists basically of a poison gland associated with a modified ovipositor. In the sterile worker and soldier castes, which do not lay eggs, the ovipositor is used only for stinging. After being jabbed into the surface of an opponent, boring movements of the lancets near its point cause the sting to penetrate deeper into the victim's body so that the secretion of the venom gland can be injected efficiently.

When a honey-bee worker attacks a mammal, the sting

apparatus cannot be withdrawn from its skin and the bee is wounded so badly that it soon dies. At first sight this appears to be a biological defect, detrimental to the survival of the species. In his famous book, *On the Origin of Species by Natural Selection*, Charles Darwin attributed the long lancet barbs of bees to an inheritance from wood-boring ancestors, such as wood-wasps and saw-flies. In these, the ovipositor of the female is used to bore holes into leaves and wood for the eggs to be deposited. It has generally been supposed that loss of the sting apparatus in the skin of a mammal was an unimportant accident – normally only insects would be stung and the sting would not be lost. More recently, however, it has been shown that the loss of the sting is no accident, but is a natural reaction developed by selection. It is normally directed against the honey-bee's largest enemies – the mammals and birds that rob the hive of its brood cells and honey.

The sting of a bee possesses a preformed breaking point, but this is weaker in worker honey-bees than in the queen. Furthermore, unlike those of the queen, the lancets of worker bees have strong barbs which anchor the sting to the skin of the victim. In consequence, the entire sting apparatus, with its nerve ganglion, is ripped out and continues working, so that every drop of its venom can be injected into the enemy. The queen bee, on the other hand, only stings rival queens, from whose bodies she can withdraw her sting unharmed. The behaviour of attacking workers also implies that their sting apparatus has been evolved for use against mammalian enemies rather than insects. They are stimulated by warm, moist breath and, possibly, by the odour of sweat. They place their stings on dark areas of hide, which are much larger in size than an insect would be, and twist their abdomens in a way that helps to tear out the sting which is much smaller and more difficult to brush off than a whole bee. Finally, the site of the sting is marked with an odorous warm substance which attracts other bees and incites them to rally there in formidable numbers. The attack of a large vertebrate endangers the survival of the entire hive, so the loss of a few dozen workers in warding off enemy attack is not a significant one to the colony.

The sting is reduced in many species of ants, but the significance of this evolutionary trend is far from clear. Some scientists have attempted to correlate feeding habits with sting reduction, suggesting that ants with effective stings are carnivores while herbivorous species have reduced stings, but this does not really correspond with the facts. The safari-ants (Dorylinae), for example, are exclusively predatory, yet they have a much reduced sting apparatus which cannot be used for killing prey. A more probable explanation is as follows: the most important enemies and prey of ants are not vertebrate animals, but arthropods – especially other ants. Because of its position at the tip of the abdomen, the sting of an ant is an awkward weapon to use against small, fast-moving insects – even though the narrow waist increases its mobility. Only the more primitive ants have retained its full use, and many of these fight with their mandibles. Of the remainder, only the smaller species are able to use their stings successfully during fights with other ants.

In place of ponderous stings, many ant species have developed chemical weapons for their defence. These include toxic secretions of various kinds which, as we have seen, can be rapidly sprayed at an enemy. Some of them are able to penetrate the waxy cuticle of an opponent, and are specific insecticides whose killing powers even exceed that of DDT. Such chemical defences have apparently influenced the morphological evolution of these ants. The armour of their integument is greatly reduced, allowing increased manoeuvrability, and greater speed and accuracy in the discharge of their repugnatorial sprays.

Defence of the colony plays a very important role in the life of social insects which cannot survive for long outside their nests. It seems possible that the reason why males take no part in the formation of hymenopteran communities may lie in the fact that they have no ovipositor or sting. Consequently, they can neither capture prey nor defend the colony. As drones, they are tolerated only until they have fulfilled their biological task of inseminating new queens; then they are cast out and destroyed. In contrast, the chemical defence weapons of nasute termites are found in both sexes. Workers and soldiers can be male or female, and every caste contains representatives of each gender.

Possession of a sting in the tail is not restricted to wasps, bees and ants. Scorpions are among the best known of poisonous animals and, by the unreasoned terror they inspire as well as by the strangeness of their form, have influenced the imagination of the peoples of the Mediterranean and of the Orient since ancient times. In Greek mythology, Orion, son of Zeus, defied Artemis, the goddess of hunting, who produced a scorpion which stung and killed him. Thereupon, Zeus turned Orion into the constellation of stars that was said to bring rain and storms. When Mithras sacrificed the bull – the first being created, whose blood was to fertilize the universe – a scorpion was sent by Ahriman, the embodiment of evil, to destroy the source of life by attacking the testicles of the bull. Scorpions are prominent in monuments to Mithras, whose cult persisted in North Africa until the third century AD. Scorpions are frequently represented on monuments in ancient Egypt, and are mentioned in several passages of the *Book of the Dead*. Numerous incantations had the reputation of affording protection from the stings of scorpions, which are also frequently mentioned in the Bible and Talmud as formidable and dangerous animals Aristotle wrote that their sting was fatal in some lands, and inoffensive in others. But it was not until 1668 that Francesco Redi made a pioneer investigation of scorpions which was founded on observation and experiment, and excluded mythology. Redi, one of the first followers of the method of Descartes, studied scorpions in Italy, Egypt and Tunisia.

Scorpions are very much maligned for, despite the large number of species known, only a few are really dangerous to man. These few are found mainly in North Africa and the Middle East, Mexico and Brazil, in desert and semi-arid environments. The diversity of species and wide distribution of scorpions throughout the tropics and sub-tropics clearly testify to the success of this kind of animal – a success which derives in no small measure from the possession of a poisonous sting.

The stinging apparatus of scorpions consists of two venom glands, each with a separate duct, leading to the pit of the sting, whence a common duct ends in a small elliptical opening on the outer side, just below the point. When a scorpion stings, it ejects a controlled amount of venom. If the prey struggles violently, it may be poisoned, but many small spiders and insects are devoured without ever being stung. Stings delivered in self-defence usually contain a maximum dose of poison. The venoms of scorpions are of two kinds: the first causes mainly a local action, rapid and transient, while the second produces not only local but primarily a general systemic reaction, especially on the

Tooth & Claw

nerve centres. The scorpion sting probably evolved, in the first instance, as a means of subduing prey: only secondly has it acquired a defensive function.

One of the many legends about scorpions is that, if surrounded by a ring of fire, they will sting themselves to death. Although this myth has, for many years, been thoroughly discredited and shown to be due to inaccurate observation, it still continues to raise its hoary old head. When a scorpion is in peril, it carries out instinctive defence reactions and, if thoroughly stimulated, will make threatening movements and even lash wildly with its tail. These actions can give an impression that the animal is stinging itself, but only very rarely has a scorpion been known actually to strike its own body and, when this has occurred, it has almost certainly been accidental. Anyhow, scorpions are resistant to the action of their own poison. Furthermore, on theoretical grounds, it can be argued that an anthropomorphic interpretation must be involved in the suicide idea since no invertebrate animals and few mammals have the power to reason. None has the mental capability to argue that rapid death might be preferable to the pain of being burned! Finally, a suicidal instinct would inevitably be eliminated by natural selection during the passage of time since it could have no biological advantage to a species while, in certain circumstances, it would be positively disadvantageous.

In *A Cure of Serpents* Alberto Denti di Pirajno described a lady magician of Tripolitania who could charm scorpions, and handled them as though they were harmless crickets. On one occasion, she placed one in her mouth. Its tail, quivering and lashing out in every direction, protruded between her lips, yet though the poisonous sting struck her chin and nostrils, it produced no harmful effect. West African pilgrims, on their way to Mecca, are much respected in the Sudan on account of their magical powers, in the demonstration of which they sometimes perform with snakes and scorpions. I believe that, in such cases, the point of the sting of the scorpion has been broken off and blunted, so that the animal is incapable of causing harm.

Occasionally, in Africa, one meets someone who will pick up scorpions with his hand, holding them loosely, but the creatures make no attempt to sting. Either the man will be wearing as a charm a piece of dried root of a plant, often hibiscus, or he will previously have drunk a decoction from the root of this plant.

Possibly he handles the scorpions without nervousness, and thereby does not irritate them, because he has complete confidence that he will not be stung. In some parts of the Sudan, scorpions are placed in sesame oil where they die and disintegrate. The oil is then kept as a remedy to be applied to the site of a sting. In other parts of the country, the afflicted area is rubbed with the charred toe-nail of a baboon, of which a dried foot is worn as a charm in readiness for an emergency.

It is another common myth that black scorpions are more venomous than yellow ones; in fact, all the really dangerous species belong to the family Buthidae, most of whose members are yellow in colour. The venom is a clear or slightly opalescent liquid which dries to a whitish powder. At least two kinds are found: the poison of some scorpions is local in effect and comparatively harmless to man while other scorpions produce a neurotoxin resembling some kinds of snake venom, and causing partial paralysis and fever. It also has a haemolytic effect, destroying red blood cells, and can be extremely dangerous.

The symptoms caused by scorpion venom of the less virulent type consist mostly of sudden, sharp pain followed by numbness and local swelling which pass away within an hour or two. Species whose poison is neurotoxic, however, produce intense local pain at the site of the sting, often without swelling or discoloration beyond a small area of gooseflesh. A feeling of tightness then develops in the throat, so that the victim tries to clear imaginary phlegm, the tongue develops a feeling of thickness and speech becomes difficult. The patient next becomes restless and there may be slight, involuntary twitching of the muscles. Small children at this stage will not be still: some even attempt to climb up the wall or the sides of their cots. A series of sneezing spasms is accompanied by the continuous flow of fluid from nose and mouth which may form a copious froth. Occasionally, the rate of heart beat is considerably increased. Convulsions follow, the arms are flailed about, and the extremities become quite blue before death occurs. This complex pattern of reactions may last from 45 minutes to 10 to 12 hours. If the victim recovers, the effects of the venom persist for longest at the original site of the sting, which may be hypersensitive for several days, so that only a slight bump will send painful or tingling sensations throughout the surrounding area.

Tooth & Claw

The enemies of scorpions include various large centipedes, spiders, camel-spiders, lizards, snakes and birds. In addition, scorpions are inveterate cannibals. When numbers are kept together in the same container, cannibalism takes a steady toll. Moreover, the addition of food in the form of cockroaches and other insects actually increases the slaughter. The cockroach is grabbed by a scorpion, two or three others rush up, and a grisly tug-of-war ensues during which, as often as not, one scorpion loses it grip on the cockroach but gains a firm hold on one of its competitors and begins eating him instead. In the battles, no holds are barred, the animals dragging with their claws and thrusting at one another with their stings, while tails frequently become entwined as they lever their adversaries from the food. In capitivity, female scorpions frequently eat their own young. I once saw a female which had recently given birth to a large family, busily engaged in a mating dance, at the same time chewing one of the babies. This must surely rank as an extreme instance of utilitarian turpitude!

At temperatures found in the tropics, scorpions strike at lightning speed, lashing at each other viciously. Their stings are seldom capable of penetrating the dorsal surface of an opponent's body, but can easily be inserted in the soft underparts. Once firmly gripped in the claws and jaws of an enemy, however, scorpions usually cease to put up any effective opposition. The tail of the unfortunate victim often continues to wave feebly for half an hour or so, while its possessor is being slowly devoured. In the tropical rain-forests of Africa and America, scorpions are sometimes caught by raiding armies of driver-ants. Although many times the size of their tormentors, scorpions rapidly succumb to their attacks and are dismembered. In contrast, giant millipedes with their integumental armour and repugnatorial chemicals are quite immune.

Baboons and other monkeys become adept at catching scorpions without getting stung, tearing off the tail and greedily devouring the rest of the body. No defence can be one hundred per cent effective: despite their stings, scorpions are extremely vulnerable to vertebrate enemies on account of their comparatively large size and slow locomotion. The long beak of a stork, for example, would snap one up without a thought, and antelope might well trample a scorpion underfoot, just as mule-deer will

Venoms

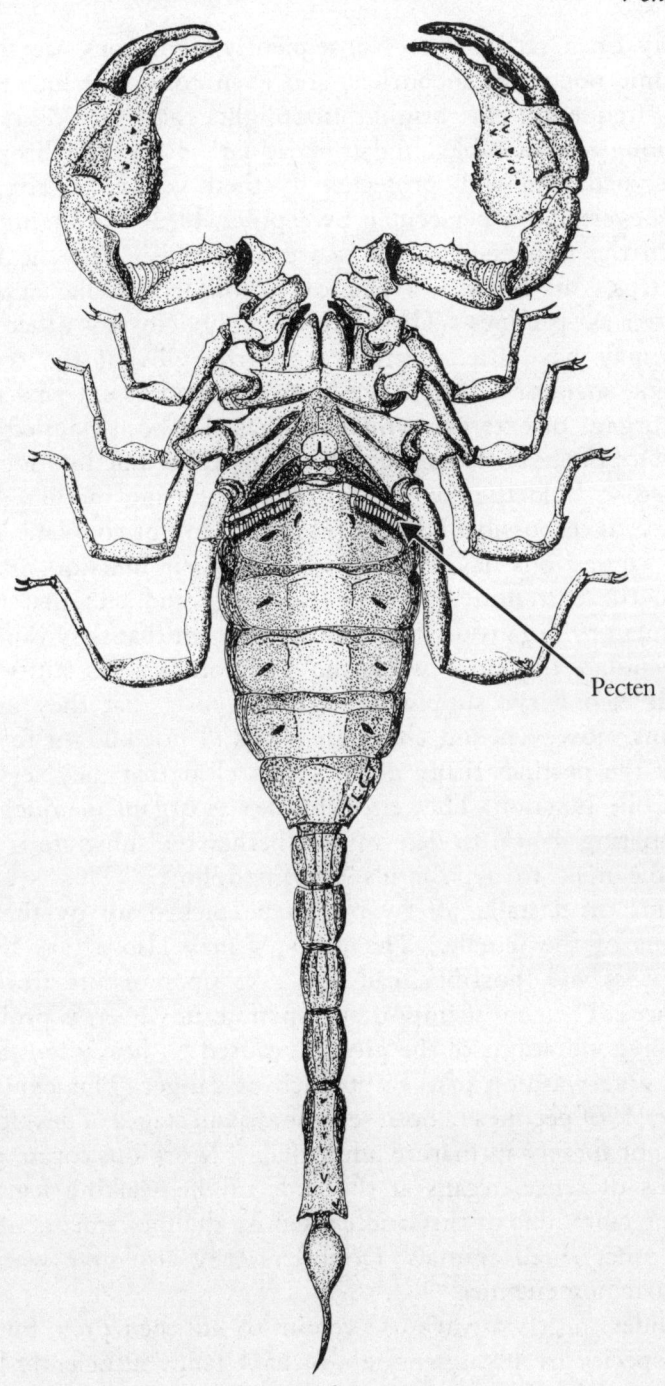

18. Ventral view of a scorpion showing the pectines.

stamp on a rattlesnake. Consequently, scorpions are forced to become nocturnal anchorites, and even come out into the open less frequently in bright moonlight than on dark nights. Although formidable predators, as far as other arthropods are concerned, and well protected by their venomous stings, they are, nevertheless, persecuted by reptiles, birds and mammals.

On the underside of the scorpion's body, near the bases of the legs, can be seen a pair of curious comb-like appendages known as 'pectines'. These are not found in any other animals and may have been derived from the gills of the scorpions' marine ancestors. They have been regarded as external respiratory organs or external genitalia: it has also been claimed that the pectines of the male and female scorpion became interlocked and served to hold the two sexes together during mating. This, of course, is impossible because scorpions do not copulate. Alternative suggestions have been made that their function may be to clean the extremities of the claws, legs and tail; that they are stimulatory organs used in mating; or even that they can be used to ventilate the lungs while the scorpion is in its stuffy retreat! Their rich nerve supply strongly suggests that they are sense organs, however, and although it is still not known for certain what the pectines really do, it seems clear that they serve more than one function. They probably act as organs of touch during the mating dance, to determine whether the substrate is suitable for the male to deposit his 'spermatophore'. (This is a bag of sperms on a stalk, afterwards to be picked up by the genital organs of the female.) The pectines may also act as humidity receptors and, possibly, can also take up moisture from damp surfaces. Their most important function, however, is probably to perceive vibrations of the ground, caused by heavy footfalls, and thus give warning of the approach of danger. This explains the presence of pectines in both sexes and at all stages of development, and not merely in mature adult males. Scorpions locate prey by means of sense organs at the ends of the walking legs. These detect vibrations of the sand caused by the movements of insects and other small animals. Doubtless they also give warning of approaching enemies.

Spiders nearly always use venom to kill their prey, but only a few species are strong enough, or have fangs sufficiently long, to penetrate the skin of a vertebrate enemy. Consequently, spiders

Venoms

tend to rely upon concealment, disguise or mimicry to avoid predation rather than on their poison fangs. Nevertheless, poisoning and the occasional human deaths do follow the bites of certain species. Some of the tropical funnel-web tarantulas can be rather irritable, but no spider bites without any provocation or disturbance whatsoever. It is curious that, even within the same genus, one species of spider may be harmful to Man while another is not. The brown recluse *(Loxosceles)* is not aggressive, and bites only in defence – as when accidentally squeezed against a human body while under clothing – while black widows and shoe-button spiders *(Latrodectus)* will crawl delicately over a human hand and then bite irritably when they find a hair too large to climb over! Female spiders are more poisonous than males, and most prone to bite when guarding their eggs or young.

Without exception, spiders venomous to Man are species in which the female takes unusual care of her young. Before laying

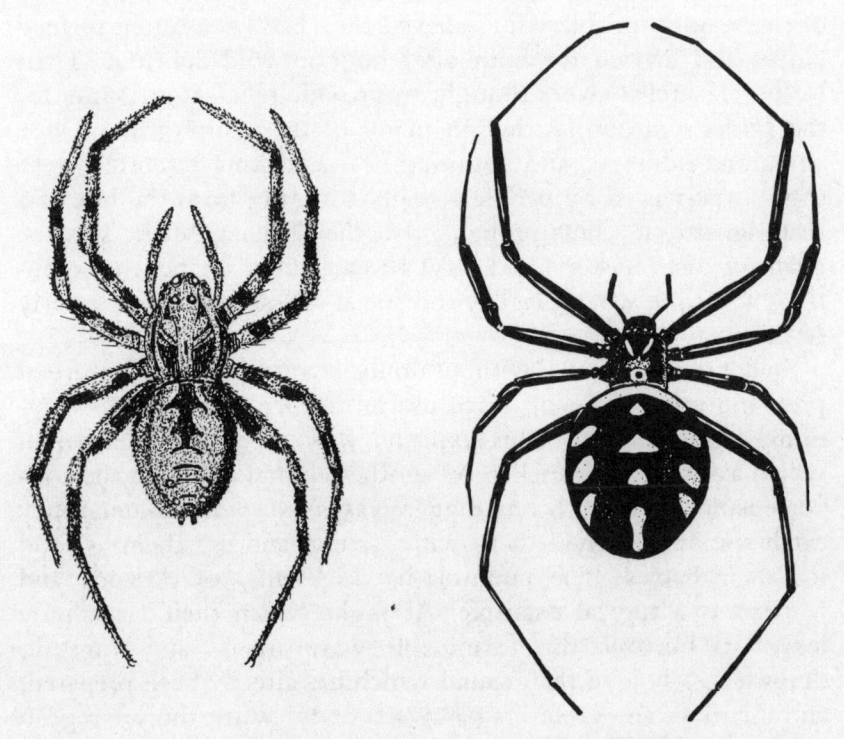

19. Tarantula and 'malmignatte'. (From J. L. Cloudsley-Thompson.)

their eggs, funnel-web spiders *(Atrax)* enlarge the lower parts of their burrows; female wolf-spiders construct underground chambers for their motherhood, and arboreal tarantulas weave nests of leaves. The eggs, secreted in silken cocoons or egg-sacs, are guarded by the mother. Wolf-spiders carry their cocoons attached to the spinnerets, while the newly-hatched young, like baby scorpions, ride on their mother's back until after the first moult. Female tarantulas carry their large cocoons between the jaws and the first pair of legs, while web-builders suspend their cocoons from silken threads.

During the Middle Ages, epidemics of hysteria, the Dancing Mania, swept across Europe and, in Italy, were believed to result from the bites of a harmless wolf-spider or 'tarantula' which acquired her name either from the city of Taranto or from the river Thara in Apulia, on the banks of which the spider was said to have been most frequently found. It is possible that the tarantula was unjustly blamed for the venomous bites of the 'malmignatte', a relation of the American black widow. To rid the body of the venom, people who believed they had been bitten formed circles and danced for hour after hour in wild delirium. Their lively 'tarantellas' were thought to provide relief, if not cure, for the spider's poison. Although many of the symptoms, such as agitation, giddiness, and vomiting bear a striking resemblance to those experienced by people actually suffering from the bite of a malmignatte, it seems probable that the Dancing Mania was the result of mass hysteria and served as an outlet for pent-up emotions at a time when plagues and social stresses were particularly severe.

Spider venoms have, without doubt, evolved for the capture of prey and are seldom of much use in defence – except in exceptional circumstances. This explains why no case is known in which a spider is mimicked by another animal although the converse is not unusual. Many digger-wasps are spider hunters: they catch spiders, paralyse them with a sting, and use them as food for their babies. The tarantula-hawk *(Pepsis)* of Arizona and Mexico is a typical example. At night, when their large prey leaves its burrow, this formidable wasp attacks and stings it, drags it to a hole in the ground which has already been prepared, and then lays an egg on its paralysed body. When the wasp grub hatches, it feeds on the motionless but living body of the unfor-

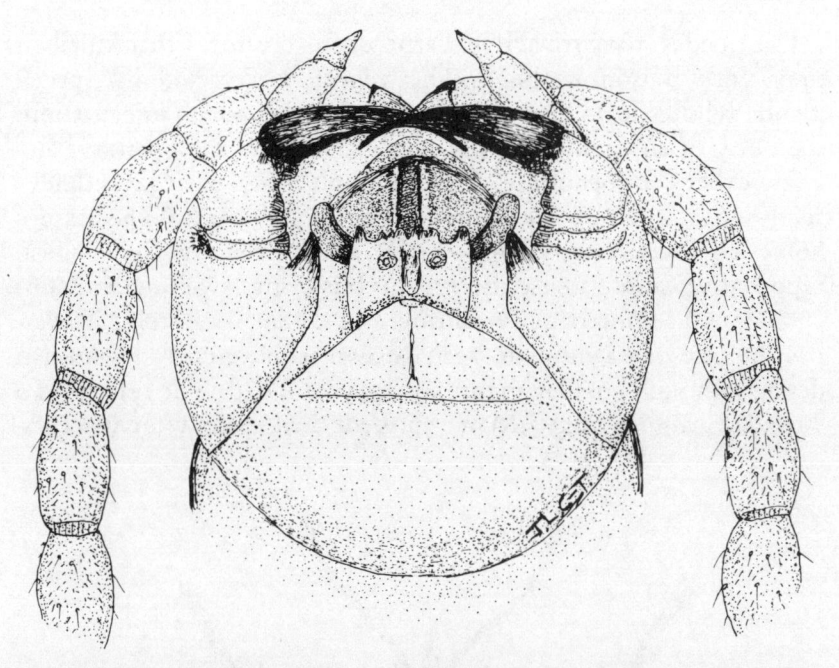

20. Ventral view of the head of a scolopendra showing poison claws.

tunate tarantula. All kinds of spider may be attacked although the wolf-spiders are, perhaps, the most frequent victims. The first action of a hunting wasp when it attacks a spider is to remove the latter from its immediate surroundings – a garden spider is much more vulnerable when torn away from its web, and a burrowing spider dragged into the open is nearly defenceless. Spiders appear completely panic-striken when confronted by a hunting wasp. Their immediate reaction is to flee, their best chance of survival, and they do not try to defend themselves with their jaws when cornered as this would probably be useless.

Poison is not found in arachnids other than scorpions, spiders and false-scorpions, although whip-scorpions and camel-spiders have sometimes been mistakenly credited with having a venomous bite. The tiny false-scorpions which inhabit leaf-litter and are found under bark have no tail or sting; but they are equipped with poison glands in their claws. These appear sometimes to be used in the capture of prey, but little is known of the enemies of false-scorpions nor of how effective their poisonous jaws can be in defence.

Tooth & Claw

Centipedes, too, are active, carnivorous creatures that kill their prey with poisonous bites. The venom is secreted by special glands whose ducts lead to the poison claws, which inject it into the body of an enemy or prey. Only the larger scolopendras and a few other centipedes are strong enough to use their jaws defensively against vertebrate animals. The larger tropical species do, however, occasionally bite Man. Their venoms have been reported to cause intense, fiery pain, blistering, swelling and subcutaneous haemorrhage, vomiting and dizziness. Rarely, if ever, has the bite of a centipede been known to have caused a human death, although smaller mammals may be killed. The refrain of a Trinidad calypso runs: 'Man centipede bad, bad; woman centi-

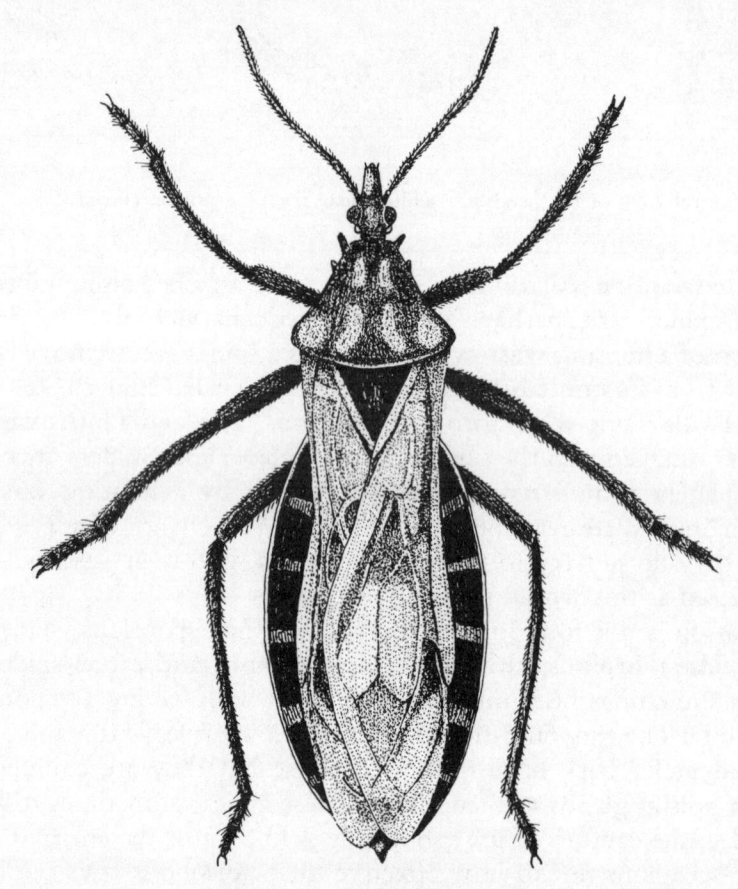

21. Assassin-bug. (From J. L. Cloudsley-Thompson.)

Venoms

pede worse than bad', but Indian children have been seen to drag huge centipedes out of the earth, and eat them!

Many insects possess salivary pumps for injecting poison into the bodies of their prey and some, such as assassin-bugs (Reduviidae), use the same mechanism in defence, as well as spraying their enemies with the poison. A few of these bugs have acquired the habit of sucking blood from vertebrates, and a couple of species transmit Chagas' disease to Man in South America. Most of them, however, are predacious, living on the blood of other insects. In defence, they will readily jab their proboscis into any animal, including man, and the bite can be extremely painful. It is believed that the habit of feeding on mammalian blood may have been acquired in the first instance from the use of the proboscis in defence. The pain has become reduced with the development of regular blood-sucking, and species that normally feed on human blood can do so without disturbing their victim, although the bites cause considerable irritation later.

In olden times, the emirs and khans of central Asia used to keep assassin-bugs of the kind whose bite is excessively painful for the express purpose of torturing prisoners. In 1842, Nuzaffer ed-Din, the Emir of Bokhara, whose own subjects said of him that he must have been suckled by a tigress, tormented in this way two British officers who had been sent to him on a diplomatic mission. The Emir flung them into his bug-pit where they suffered the most ghastly agonies for several months before he hauled them out and beheaded them in the market place!

While, at first glance, the jellyfishes, sea-anemones and corals may appear to have little in common with one another they are, in fact, members of the same phylum – Coelenterata or Cnidaria – containing animals alike in being radially symmetrical and possessing tentacles armed with stinging structures called 'nematocysts'. Coelenterates are of two basic structural types, polyps and medusae, both of which may be present in the life cycle of a single species. The Portuguese man-of-war *(Physalia)*, one of the most dangerous, consists of a bluish medusoid float, 10 to 30 cm long, with tentacles up to several metres in length, attached to which are numerous polyps armed with nematocysts whose sting causes painful local reactions, swelling of nearby glands and generalized symptoms. *Physalia* has been reputed, on occasion, to cause the death of human victims, but the most dangerous of jellyfish and probably the most venomous marine organism known is the sea-wasp *(Chironex)* whose sting can kill a man within 3 to 8 minutes.

Contact with the tentacles of those coelenterates whose nematocysts are capable of penetrating the human skin may result in symptoms ranging from an immediate mild, prickly or stinging sensation, like that caused by the sting of a nettle, to an intense burning, throbbing or shooting pain which may even render the human victim unconscious. In some cases, the pain is restricted to an area within the immediate vicinity of the contact; in others, it may radiate to the groin, abdomen or armpit, depending upon the site of the sting, or become generalized. The localized pains may be followed by a feeling of numbness. The skin develops a rash, similar to that of urticaria, with blisters and swelling. Later, multiple abscesses, necrosis and sloughing of the tissues may develop. The stings of jellyfishes in particular cause redness and flushing of the face, perspiration, lacrymation, coughing and sneezing.

Coelenterates are carnivorous in their feeding habits. Some of the bigger jellyfishes capture and devour large crustaceans and fishes, while the food of anemones consists of molluscs, starfishes and brittle-stars, crustaceans and fishes. Coral polyps are so small that they can only eat small crustaceans. Nevertheless, milleporine corals are not only protected by their skeletal armour, but have powerful nematocysts which readily penetrate the human skin. Although primarily offensive weapons, evolved for the capture of prey, nematocysts are also used for defence; and most other marine animals keep a respectful distance from coelenterates if they can.

Considering their minute size, which ranges from 5 μg to a length of 1.12 mm, the structure of nematocysts is remarkably complex. They consist of a thread or barb, with folded spines, coiled inside a capsule containing venom: the capsule is closed by a lid or 'operculum'. Projecting from the outside of the capsule is a trigger-like 'cnidocil'. Nematocysts are sensitized by quite infinitesimal amounts of a protein, glutathione, which is produced by most animal tissues, and are fired through contact with the cnidocil. The operculum opens, the barb absorbs water instantaneously, swells, turns inside out, shoots through the opening of the capsule and enters the flesh of the animal that

Venoms

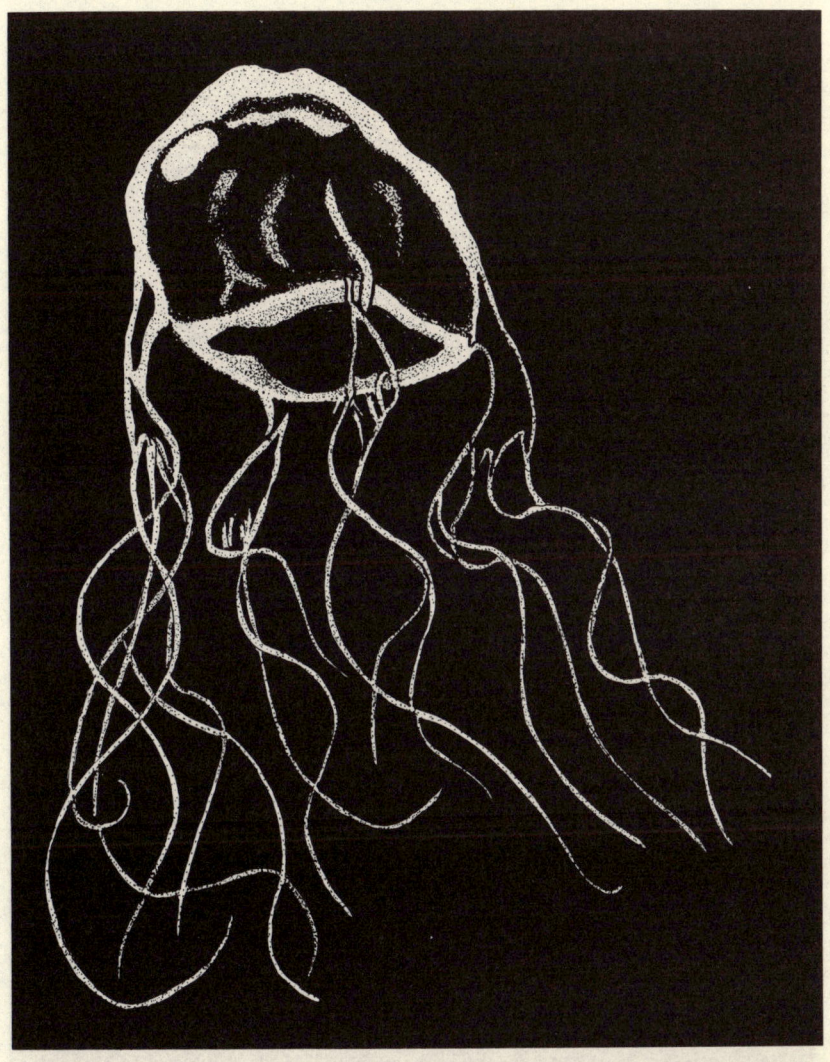

22. The sea-wasp *(Chironex)*, most dangerously poisonous of jellyfishes. (After W. Bücherl and E. E. Buckley.)

discharged it. It is quite extraordinary that sea-slugs not only manage to feed on sea-anemones without discharging the nematocysts, but actually use them for defence – as we shall see in the next chapter.

Venomous species of molluscs are only found among the carnivorous marine snails and the cephalopods – squids, cuttlefish

117

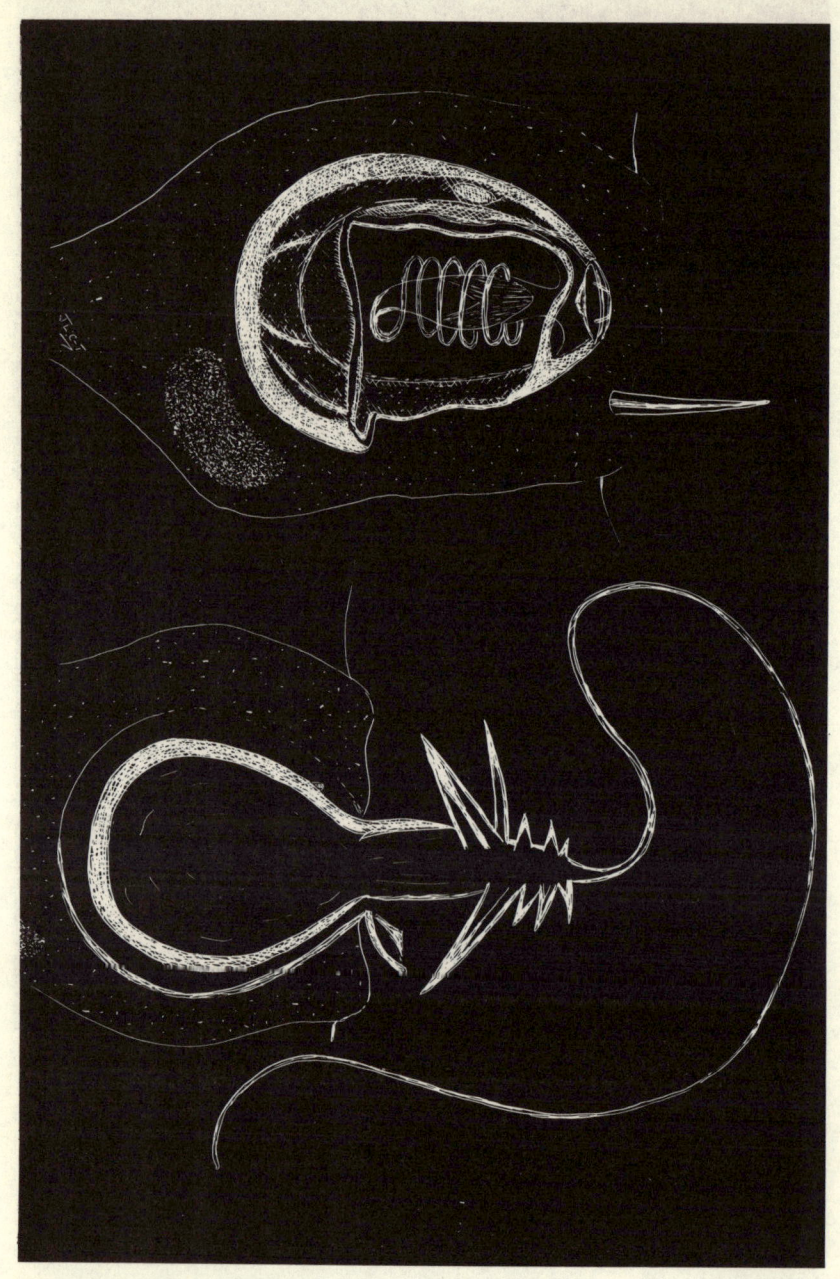

23. Stinging cells (nematocysts) of a sea-anemone. *Above*, undischarged; *below*, discharged. (After W. Bücherl and E. E. Buckley.)

and octopuses – which also use poison to kill their prey. The use of poison in defence is secondary to its offensive function. The primary defence of snails lies in the shell; cephalopods usually avoid their enemies by swimming away from them. Of four families of venomous marine snails, only Conidae or cone-shells are known to be actively venomous to man. Among the Cephalopoda, venom is usually present for the capture of prey, but only the octopods have so far inflicted poisonous bites on human beings.

Of the species of cone-shell known to be responsible for stinging Man, the geography-cone *(C. geographicus)* is the most dangerous. It is known to have been responsible for a number of deaths. Like many other poisonous snails, its natural food is fish, but even cones that feed on worms and molluscs can cause serious harm to susceptible individuals who allow themselves to be stung. The venom apparatus in cone-shells consists of a single poison gland and duct, a storage sac in which are housed a number of modified radular teeth, into which the duct of the poison gland opens, and a retractable proboscis which carries the teeth forward and thrusts or shoots them into the prey. The nature of the venom is little understood: it acts directly on the muscles, causing paralysis. The symptoms in Man consist of a sharp sting and numbness, bluish discoloration of the skin at the site of the puncture, paralysis spreading to all parts of the body including the throat so that speech is affected, loss of sight and consciousness, followed, in severe cases, by death.

The bites of some octopuses cause local effects, like those of a bee sting, but other species can produce serious symptoms in Man. A curious feature is the complete absence of pain at the time of the bite. Shortly afterwards, however, the victim experiences dryness of the mouth, numbness of the lips and tongue, nausea and difficulty in swallowing, followed by numbness spreading to all parts of the body, vomiting, failing eyesight, loss of muscular co-ordination and paralysis, leading in exceptional cases to death. The chemistry of the venom is not yet known, but it may well be similar to that of cone-shells. When injected into corals, fish and other prey, it causes rapid paralysis.

Venoms are rare among higher vertebrates: there are, however, plenty of fishes with poisonous spines, a few with venomous bites, and several amphibians with poison glands in their skin.

Tooth & Claw

Two species of lizards, and many snakes, have poisonous bites, but there are no venomous birds; while, apart from the poison spurs of the platypus and echidnas, only shrews and solenodons among mammals have a poisonous bite. Solenodons are rare, squirrel-sized insectivores from the West Indies. Two species are known – the agouta of Haiti and the almiqui of Cuba. The snout is long and trunk-like in each, the body has bristly hair, while the tail, which is as long as the body, is scaly and with few hairs. These curious animals are slow and so awkward that they stumble over their own feet when trying to escape capture. But their saliva contains a toxin which endows them with a poisonous bite

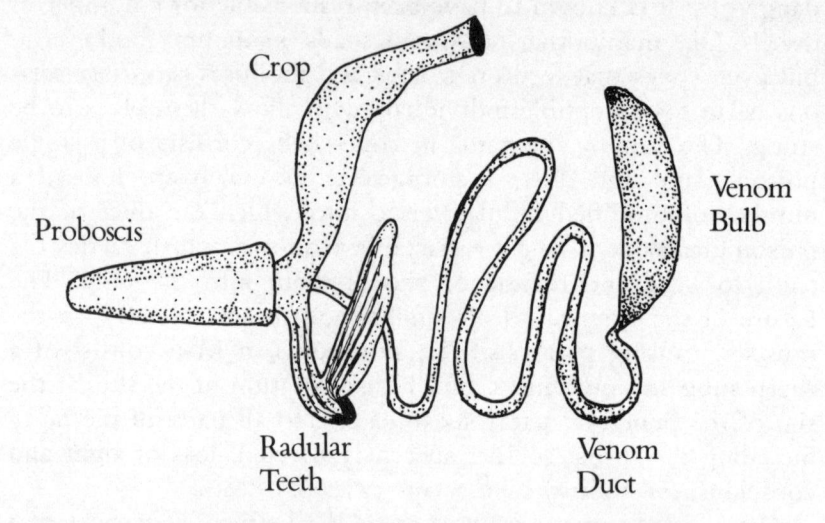

24. The venom apparatus of a marine cone-shell, most poisonous of molluscs. (After W. Bücherl and E. E. Buckley.)

and it is almost entirely on this that they rely for defence. That shrews, also, can give poisonous bites has long been known. It was discussed by Edward Topsell in the *History of Four-footed Beasts and Serpents*, published in 1607, but subsequently came to be regarded as a myth. In recent years, however, histologists studying the structure of certain cells in the salivary glands of the North American short-tailed shrew have found that the sub-maxillaries secrete a lethal poison that quickly paralyses mice and other small animals.

Without question, snakes have developed the use of lethal

poisons to a far greater extent than any other animals. Nevertheless, it is not always realized that the majority of snakes are non-poisonous. In fact, only three groups are venomous – the back-fanged snakes, the deadly cobras, kraits and mambas, and the vipers and rattlesnakes. The back-fanged snakes have to grip their prey before the poison, which usually has a haemorrhagic effect, can be chewed into the wound. It is as difficult for them to bite in self-defence, as it is for the Gila monster of Arizona, one of the two venomous lizards known – the other is a closely related species, the Mexican bearded lizard of the Sonoran desert. Their venom glands lie in the lower jaw and consist of three or four lobes, each with a duct opening at the base of one of the large mandibular teeth. These have deep grooves, flanked by sharp flanges, down which the poison flows. Its effects on man include severe local pain and swelling, accompanied by systemic poisoning of varying severity which has been known occasionally to result in death.

The venoms of poisonous snakes are secreted by modified salivary glands. They are stored in sacs with tubes leading to the fangs. The venom teeth of back-fanged snakes are merely grooved, but the fangs of vipers, cobras and other dangerously poisonous snakes are tubular, so that the venom can only escape from a minute hole near the tip. In cobras, mambas and their allies, the two large poison fangs in the top jaw are permanently erect, whereas the poison fangs of vipers and rattlesnakes are erected only in the act of biting. Normally they are folded back and concealed in fleshy tissue. In most cases a snake strikes rapidly, and disappears before it can be identified. The characteristic punctures made by the teeth, however, often enable diagnosis to be made. Where human beings are concerned this may be important in deciding what kind of treatment should be given.

Non-poisonous, and back-fanged snakes which are usually only mildly poisonous, have no large fangs, and their bites leave only double rows of small tooth marks. The presence of two large bleeding fang marks, on the other hand, is clear indication that the victim has been attacked by a dangerously poisonous snake. Vipers leave fang marks only, whereas the bites of cobras and mambas leave a row of tooth marks in addition to the deep punctures caused by the poisonous fangs. Back-fanged snakes include the cat-snakes *(Telescopus)* of North Africa and the Middle East, the hissing sand-snakes *(Psammophis)* of Africa and Asia, the South African skaapstekers *(Psammophylax)* and the lyre-snakes *(Trimorphodon)* of North America.

Cobras, kraits and mambas (Elapidae) show a very great diversity of size and colour. The Egyptian cobra, for instance, has dull, lustreless scales. It is widely distributed throughout Africa, although it seems to prefer hot, dry and sandy places where its colour is in keeping with that of the dusty surroundings. The 'asp' which Cleopatra applied to her breast, after having 'pursued conclusions infinite of easy ways to die', was either a snake of this species, or else one of the horned vipers – to which the name 'asp' has also been applied. The Egyptian cobra is often seen performing with snake charmers from Morocco to Egypt. It is a quick, irritable reptile which rears up at the slightest disturbance and strikes repeatedly, with sharp hisses.

Cobras *(Naja)* are rather active snakes which move about mainly at night, although they are also to be seen during the day. Contrary to popular opinion, they do not usually look for trouble and, if given an opportunity, will rapidly escape. The Indian cobra sometimes enters buildings and is said to be most aggressive at night. This species is widely distributed throughout South East Asia. In Malaya it is often black but, in other regions, may be much lighter in colour. It is distinguished by the characteristic 'spectacle' markings on the 'hood'. The cobra's hood is produced by the snake widening and flattening its neck by extending the anterior vertebrae in threat. The African black-necked cobra not only bites, but has the unpleasant habit of spitting venom, with considerable accuracy, at the eyes of anyone who disturbs it. This causes intense pain and, if not treated immediately, invariably results in blindness. Another spitting cobra is the smaller South African ringhals, so called because it has a whitish ring across its throat. Some Indian cobras have also developed the technique of spitting venom.

Unlike true cobras, which feed mainly on rats and other small mammals, the king cobra *(Ophiophagus)* is a snake eater. It is the largest and most dangerous serpent in the world, at times reaching a length of 6 metres. Although stories of its aggressive nature are often exaggerated, as previously mentioned it has not infrequently been known to kill elephants working in the forests of Burma and Thailand, and will boldly attack anyone or anything that disturbs or interferes with it, especially in the breeding season. The coloration of the king cobra is olive or pale brown, often with rings or bands of black. Although its anterior ribs are somewhat elongated, it cannot spread its 'hood' to the same extent as true cobras can, nor does it rear up in such a dramatic way. Occasionally, however, it will raise its head a metre or more and stay motionless, like a great candlestick, staring fixedly and without any of the nervous swaying of the Indian cobra.

Kraits *(Bungarus)* belong to the same family of poisonous snakes as the cobras. They can be recognized by the fact that the scales of the longitudinal row along the centre of the back are enlarged and there is no 'hood'. Kraits move about and hunt almost entirely at night. Like king cobras, they feed mainly on other snakes, but they may also take small rodents and other mammals. The majority of kraits are timid and reluctant to bite, even if teased. This is fortunate as krait venom is extremely toxic to Man. The mortality rate from krait bites is over 75 per cent, as compared with a death rate of under 10 per cent from cobra bites. The Chinese banded krait ranks third as a cause of snake bite in Taiwan, the pit-viper *(Agkistrodon)* coming first in this respect. Mambas *(Dendraspis)* are unique among really deadly snakes on account of their extreme slenderness. These innocent-looking snakes may be green or black, the latter variety attaining a length of up to 4 metres. The fangs are situated at the very front of the mouth and point downwards: the toxicity of mamba poison is extremely high.

Coral-snakes, so called on account of their brilliant hues, are also relatives of the cobras, kraits and mambas. They are conspicuously marked with transverse red, black and yellow bands, and occur in a variety of environments from tropical forest to savanna and semi-desert. Asian species, although venomous, are of little medical importance but, in the New World, some of the coral-snakes are very poisonous. As we shall see in Chapter 9, they may serve as models for various harmless, or mildly poisonous, back-fanged snakes that mimic them.

Sea-snakes have been referred to as 'sea-going cobras' and the epithet is quite appropriate since, like cobras and kraits, they have erect fangs and neurotoxic venoms. Most sea-snakes are marine, although a few may be found in brackish water. They have a distinctive characteristic in that the tail is flattened like the blade of an oar. Many species never leave the water but some come ashore to lay their eggs. A few attain a length of 3 metres but the majority are smaller than this. These snakes seldom bite, even when roughly handled but, when accidents do occur, they may end fatally.

Vipers fall into two separate groups, the Old World vipers (Viperinae) and the pit-vipers and rattlesnakes (Crotalinae). The former includes the horned viper, puff-adder, Russell's viper, gaboon-viper, and the European adder, as well as the saw-scaled or carpet-viper *(Echis)*. This is the 'dusty brown snakeling that lies for choice on the dusty earth; and his bite is as dangerous as the cobra's', which Rudyard Kipling mistakenly called 'Karait' in the story of Rikki-Tikki-Tavi – although it is clear from his vivid description to which species of snake Kipling was actually referring. The kraits are much larger reptiles than the carpet-viper, reaching lengths of a metre or more, and rather like cobras in general appearance.

Adders are found throughout Europe and Asia, and occur at elevations of up to 1,500 metres or more on mountains. In winter, they return regularly to specific dens where they congregate in hibernation. Russell's viper ranks with the Indian cobra as one of the most dreaded snakes of the Orient. Adults reach a length of almost 2 metres and the mortality rate among persons bitten is nearly 12 per cent. This distinctively marked snake is found in a variety of habitats, but most often in open country. It sometimes enters buildings in search of the rodents on which it feeds. In Africa, the sinister puff-adder and the hideous gaboon-viper *(Bitis)* also secrete venoms of extreme toxicity. Most vipers have thick bodies and flattened heads: their poison fangs are proportionately longer than those of any other family of snake.

Pit-vipers are so-called because they have sensory depressions between the eyes and the nostrils. These are provided with an elaborate supply of nerves and blood vessels, and act as directional receptors sensitive to infra-red heat. With their aid it is possible for a rattlesnake to strike at warm-blooded prey, even in complete darkness, at a distance of up to 45 cm or more. The victim rarely dies immediately, however, and the snake may need to follow the stricken creature for some distance before it drops. Rattlesnakes *(Crotalus)* trail their victims by means of a specialized receptor known as 'Jacobson's organ'. This comprises a pair of internal cavities at each side of the snout, with ducts leading to an opening in the roof of the mouth. Odorous particles, picked up from the air or the ground by the forked tongue, are transferred to these openings. With the exception of the eastern copperhead, the water-moccasin and coral-snakes, the poisonous snakes of North America are all rattlesnakes – of which there are many species, especially in the southern states. In Central and South America, rattlesnakes are replaced by another type of pit-viper, the lance-head, of which the West Indian fer-de-lance, the Brazilian jararaca *(Bothrops)*, and the giant bushmaster *(Lachesis)* are notorious examples.

Snake venoms are usually pale yellowish, slightly viscous fluids consisting of toxins, enzymes and other components. Whilst the injuries to Man produced by back-fanged snakes are seldom severe, the African boomslang *(Dispholidus)* is one of the few species to have caused human fatalities. Viperine and elapid venoms are far more dangerous, but it is questionable which is the more lethal, for the physiological effects of the two are different. The former causes collapse and heart failure, depending on the amount injected; the latter is neurotoxic and causes paralysis. Viperine poison usually includes an ingredient that causes clotting of the blood, while elapid venom often contains an anticoagulent and, at the same time, causes haemolysis or breakdown of the red blood corpuscles. Since snake poisons contain large numbers of toxic compounds and vary from one species to another the general picture is by no means a simple one. Neurotoxins may also act on the blood system, while blood poisons can have side-effects on the nervous system as well.

Cobra venom acts rapidly and, if death from respiratory failure does not occur within about twelve hours, the patient usually recovers quickly. In the case of viperine snake bites, death is usually less rapid; but it may be several days before the victim is out of danger, because late complications, such as haemorrhage and septicaemia, are not infrequent. This suggests that Cleopatra's 'asp' was more likely to have been a cobra than a viper.

Most victims of cobra bite experience severe pain immediately afterwards in the region of the injury, which becomes swollen and discoloured. The neurotoxic effects set in later, with feelings of drowsiness, paralysis of the face muscles, drooping of the eyelids, and impairment of speech and swallowing. The venoms of sea-snakes also produce neurotoxic effects, muscle spasm, and the passage of reddish-brown urine, due to the liberation into the blood of pigments from the injured muscle fibres. The kidneys are also severely affected, so that death may be due both to respiratory and to renal failure. The grave effects of viper bites include severe damage around the site of the injury, produced by the action of the poison on the cells of the body. Large areas of tissue may become gangrenous and be sloughed away, leaving permanent injuries – if death does not follow from heart failure or the cessation of respiration.

Despite their poison, venomous snakes are by no means immune to predatory enemies. Birds, such as shrikes, crows, storks, hornbills and hawks, have been known to attack them and, in England, the buzzard is said to prey on adders. The African secretary-bird is renowned as an enemy of snakes, which it kills with blows from its long, scaly legs, while the road-runner of North America, a handsome black and amber bird, is even able to cope with rattlesnakes in the same way. Hedgehogs sometimes feed on reptiles, including adders. When attacking a snake, the hedgehog erects the long bristles on its forehead and repeatedly bites the snake until it dies. As a rule the hedgehog's spines prevent effective retaliation from the snake while, even if it does get bitten, it is partially immune to adder's poison and seldom dies. The Indian mongoose, too, is well known as an enemy of venomous snakes, especially cobras to whose poison it has a natural immunity. On the other hand, it is no less susceptible to viper poison than other mammals of comparable size, and has to depend entirely upon its speed, agility and thick fur for protection when attacking them.

Both in Egypt and the Orient, the ability of mongooses to kill venomous snakes has attracted attention since earliest times. In his *Historia Animalia* Aristotle (*c.* 384–322 BC) wrote: 'The Egyptian ichneumon, when it sees the serpent called the asp, does not attack it until it has called in other ichneumons to help; to meet the blows and bites of their enemy the assailants beplaster themselves with mud, by first soaking in the river and then rolling on the ground.' Three and a half centuries later, Strabo said that the ichneumon killed the asp by seizing it by the tail or head and dragging it into the river, while Aelian claimed that, if no mud was near, the mongoose rolled in the sand before engaging in battle with the cobra! Pliny wrote of the Egyptian cobra: 'It is impossible to declare whether Nature has engendered evils or remedies most bountifully. In the first place she has bestowed on this accursed creature dim eyes . . . and in the next place, she has given it war to the death with the ichneumon.' Anyone who has witnessed a fight between an Indian mongoose and a snake will have been impressed by the intense concentration of the little mongoose from the moment it sees its adversary, its aggressive pugnacity, and the incredible rapidity of its movements.

Although the venom glands of snakes have been derived from salivary glands, it is not necessary for the prey to have been poisoned for it to be digested. Snakes force-fed with rats or birds can digest them quite adequately. Venom may assist active digestion, but it is by no means essential in the case of snakes, as it is to spiders and centipedes. Indeed, in some snakes it seems that poison has become far more important in defence than it is in offence.

Many poisonous snakes show conspicuous warning coloration, clearly indicating that their venom plays an important defensive role. The display of warning colours, as we shall see in the following chapter, often saves them, and other venomous animals, from having to bite or sting in self-defence, and thereby enables them to be economical in the use of poison.

8 Warning and Threat

Many venomous animals have striking coloration which makes them extremely conspicuous. 'Aposematic', or warning, coloration is of value to them because it draws attention to the fact that they possess an effective means of defence, and are best avoided. It pays such animals to advertise themselves for, if they are seen and recognized *before* an attack has been launched, they will be left severely alone.

Of course, not all conspicuous colours are aposematic: some brightly coloured animals may be cryptic against brightly coloured backgrounds – for instance, red molluscs can be quite inconspicuous when crawling on red coral. Some use conspicuous colours in territorial and courtship displays directed towards members of their own species, while others 'mimic' poisonous or distasteful species and obtain protection thereby.

The natural function of an animal's coloration cannot be considered except in relation to the habitat it normally occupies. For instance, the black-and-white stripes of the colobus monkey may seem very conspicuous but, in the high forests of East Africa, Ethiopia and the southern Sudan, where the trees have black trunks and branches draped with long grey manes of lichen and moss, these monkeys are often almost impossible to detect, even when they can easily be heard in the tree tops. The bold design of the African python, apparently so conspicuous in zoos and museums, is, in fact, a disruptive pattern which renders the snake inconspicuous in its normal environment of dark shade and sun flecks, green and yellow leaves, black bark and grey lichen. Coral fishes are among the most brilliantly coloured of all animals yet, in the normal environment, their bright hues and bold designs are generally cryptic.

Conspicuousness is often achieved by the use of vivid, contrasting colours and bold patterns. Wasps can immediately be recognized by their bands of black and yellow – the colours usually employed to attract attention to road signs and important notices. Black, white, yellow and orange, in simple but eye-catching designs, are used for advertisement throughout the animal kingdom. Aposematic day-active animals tend to be black with patterns of light colour, while harmful nocturnal species are more often white with patterns of black.

Alfred Russel Wallace was particularly impressed by the larva of a South American sphinx-moth *(Pseudosphinx)*, as broad as a man's thumb and nearly twice as long, but velvet black with circles of bright yellow and orange-coloured head, legs and tail. When approached, this gaudy caterpillar lifts the front end of its body and lashes it back and forth in a most conspicuous manner, which is further complimented by erratic waggling of the long black tail. Furthermore, the species is gregarious, and whole trees may be decorated with hundreds of the ostentatious creatures, all lashing simultaneously. Wallace realized that *Pseudosphinx* must be unpalatable and that its colour and behaviour act as a warning to potential predators – an assumption that has subsequently been confirmed experimentally.

A conspicuous colour scheme with bold patterns needs to be combined with free exposure and sluggish behaviour if it is to be effective. An advertisement would be of little value if it were concealed from view, or moved too rapidly to be appreciated. On the other hand, moderately slow movement draws attention to it. Consequently, it comes as no surprise to discover that aposematic animals usually combine sluggish movement with almost indecent exposure. The Gila monster and Mexican bearded lizard provide striking contrast to most other desert lizards, which depend for safety upon cryptic coloration, alertness, speed and burrowing habits. Not only are the venomous species brightly coloured – black and pink or black and yellow respectively – but they are as ill-equipped for speed as they are defiant and formidable in defence. Similarly, aposematic frogs and salamanders are not only slow in their movements but make no attempt to escape when captured, and the same is true of most warningly coloured insects, whether they have poisonous stings, repugnatorial glands or defensive sprays.

Puff-adders are awfully slow in getting out of the way when

sunning themselves on a sandy path, so that people, not looking where they walk, sometimes tread on them accidentally and get bitten. My wife and I came across one at Nimule, in the southern Sudan, which refused to move until I threw a pebble at it. Many years ago, I was at a picnic near Pretoria when one of the girls in the party, dressed only in her bathing costume, stepped clean over a sleepy puff-adder without even noticing it! On another occasion, a well-known zoologist friend had a horrid experience on Mount Cameroon. While walking along an overgrown track, he felt something squirming underfoot. Looking down, he found himself standing on the tail of an immense gaboon-viper. Luckily the snake was concentrating only on escape and my friend, who found himself sharing the same ambition, lifted his foot and hurriedly departed in the opposite direction!

The function of warning coloration lies in the fact that, if potential enemies are visually repelled, there is no need for an animal to waste its energy in poisoning them or driving them away. Nature is always economical. Many distasteful arthropods compensate for their lack of size by the habit of living in dense and often conspicuous aggregations. Instead of spacing themselves more or less uniformly throughout their environments, they occur in distinct sporadically distributed clusters. This is true of brightly coloured saw-fly larvae, blister-beetles and their allies, and a variety of bugs (Hemiptera). In some cases, aggregations are maintained by the release of volatile pheromones – chemicals used for communication within a species – which attract both males and females – a subject to which we shall return in later chapters. Evil-smelling pentatomid bugs are frequently found crowed together on leaves, where they form conspicuous patches. The same is true of gregarious saw-fly larvae and poisonous moth caterpillars which often aggregate in glaringly conspicuous splashes of colour.

The disagreeable stinging cells of sea-anemones, as already mentioned, are put to good use by the sea-slugs which feed on them. In some mysterious way, these molluscs are able to eat anemones without discharging their nematocysts. These poisonous weapons are transferred inside the body to the papillae or 'cerata' which clothe the backs of the sea-slugs, and here they provide effective defences – of special importance in the absence of a shell. So effective are the batteries of stinging cells acquired

Warning and Threat

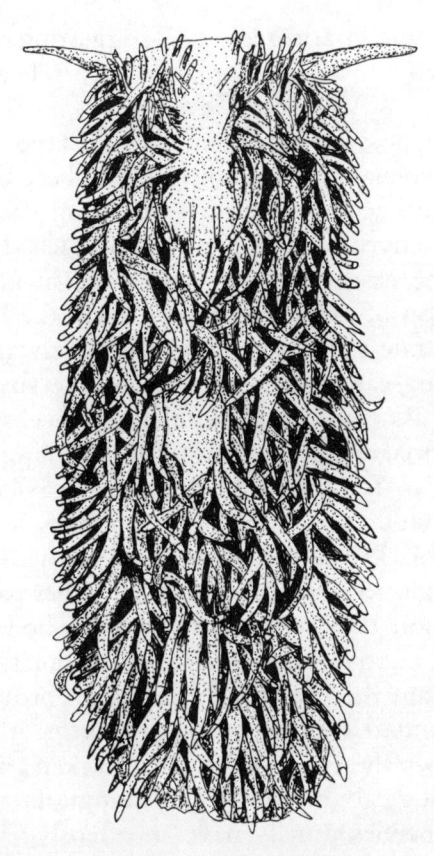

25. A sea-slug.

from their food, that starving fishes in the laboratory have even been known to eat shelled molluscs which had been pickled for years in formalin, in preference to living sea-slugs! The formidable defences of sea-slugs are advertised by brilliant orange and other conspicuous colours which appear particularly brightly on the papillae.

Those desert animals that do not possess concealing coloration are usually black, a fact whose significance has been the subject of much dispute among biologists. An apparent paradox is presented by the fact that dark colours generally absorb more heat than paler hues. In some cases, black pigmentation may be an evolutionary legacy which has been retained because it is not directly disadvantageous. This explanation may apply to black desert-beetles, crows and other birds, but the proportion of species and individuals which are black, especially among insects and birds, increases in arid regions – suggesting that some positive adaptive value to the colour or to a correlate of it, must be implied.

There are two possible explanations: either the black colours of desert animals have a thermal significance – because they enhance heat gain in the evening and early morning they enable small animals to be active for a longer period each day than would otherwise be the case – or else their function lies in advertisement. Absorption of radiation cannot warm the body of an animal, however, unless the heat is conducted inwards. The cavity beneath the wing-cases of desert-beetles, the space beneath the feathers of birds, and other insulating mechanisms militate against this. Moreover, if the black exterior should become very much hotter than the environment, the excess heat would be removed by conduction, convection and, to some extent, by radiation. The black insects that melt icy graves for themselves on the edges of glaciers and Arctic snow patches pass on to the ice the solar radiation they absorb. Their own body temperatures cannot rise much above zero. Moreover, colour is probably relatively unimportant thermally because a high proportion of solar energy is transmitted at infra-red wave lengths. It must therefore be concluded that the functions of the black pigments of desert animals are ecological rather than physiological.

Black desert-beetles usually have an offensive smell and taste. If crowded together in a bottle, or some other container, with inadequate ventilation, their fumes will not only kill any other insects placed with them inside the bottle, but even results in the death of the beetles that have secreted them. Many other desert insects are also intensely black in colour and, although non-toxic, most of them have extremely hard integuments, or may be distasteful in other ways.

For a yellow wasp to become conspicuous and possess warning coloration, it must acquire black markings. Similarly, on a yellow background of desert sand, black is by far the most conspicuous colour; reds, yellows or browns would not show up at all well. Consequently, it is natural to expect any conspicuous desert animal to be black, as indeed they nearly all are. The functions of such advertisement may be diverse – warning, social interaction, and so on. There are many positive advantages, as well as disadvantages, in being conspicuous.

Warning and Threat

Animals with cryptic coloration are only inconspicuous while they are motionless, and they usually 'freeze' at the approach of danger. In contrast, animals with aposematic coloration show the opposite tendency – that is, they have an instinct to move about and make a noise on the approach of danger. This can be seen among poisonous snakes which make an exhibition of themselves when driven to extremities – gaping, hissing, striking wildly, or drumming on the ground with their tails. When angered or in danger, agamid lizards jerk their heads up and down. This makes them extremely conspicuous to a rival or to an enemy, and may be effective, both in territorial disputes between males and in scaring other small animals. If the case of larger predators, however, the male agamid's behaviour places him in great danger. Indeed, this may even be of benefit to the species. If the conspicuous males are sacrificed, females, which are cryptic, may have an increased chance of survival: and females, of course, are of greater significance in the furture of the species.

Conspicuousness can be achieved by warning sounds as well as aposematic colours. One of two kinds of mechanism is usually responsible for the production of sound in the animal kingdom: the vibration of elastic structures in the breathing tube, and stridulation caused by the friction of rigid parts. The sounds produced by vertebrates come within the first category while most invertebrates use methods belonging to the second. The stridulatory apparatus consists of two parts, a file or 'strigil' on the animal's body which is rubbed with a scraper or 'plectrum'. The file may be composed of hairs, spines, tubercles, teeth or ridges, while the scraper is a projection with a sharp edge, or the tapering edge of an appendage. The distinction between these two components of the apparatus is generally convenient but somewhat artificial, and at times it is not possible to differentiate between them. In ants and mutillid wasps, for instance, the file and scraper are situated on different abdominal segments, in grasshoppers on the wings and legs while, in crabs and lobsters, the stridulatory organs may be on the claws, cephalothorax or tail. The buzzing of wasps and bees and the drone of flies is probably caused by the vibration of the wings and thoracic muscles, and by the expulsion of air through the respiratory apertures or spiracles.

Tooth & Claw

The large death's-head hawk-moth *(Acherontia)* is distasteful and not only has conspicuous markings, but is unique in being able to produce a squeak by the forceful expulsion of air through its proboscis. Other distasteful Lepidoptera make chirping noises by emitting repugnatorial foamy matter which flows through orifices on the top and sides of the thorax. Liquid discharged from the abdominal defence glands of brachynid beetles contains, among other things, nitric acid and a lipid. Ejected outside the body, it explodes, producing an audible detonation, and volatilizes into a white vapour, which is phosphorescent in the dark.

26. Death's-head hawk-moth *(Acherontia)*.

The male cicada possesses one of the most complicated sound-producing organs in the animal kingdom. He 'sings' by rapidly vibrating a pair of membranes or 'tymbals' situated on the abdominal segments of his body. These are pulled inwards by special muscles and regain their former shape by means of their natural elasticity. The mechanism is the same as when a rounded tin lid is pressed by the finger, only to regain its former shape accompanied by a sharp click. In some species the song is quite deafening and becomes exceedingly monotonous. From afar it sounds like a cross between a distant threshing machine and a frog pond. In others it has been likened to a hand-bell, a steel knife on a grindstone, a steam whistle or the rattle of a rattlesnake. But its function is sexual, not aposematic: female cicadas do not possess sound-producing organs, and the satirical poet Xenarchus wrote, 'Happy the cicadas' lives, for they all have voiceless wives' – a couplet that has since been quoted *ad nauseam*!

The stridulation of small spiders is, no doubt, also concerned with mating, but larger tarantulas stridulate defensively by rubbing their jaws together. The friction of the two appendages, whose inner faces are studded with club-shaped bristles, produces a clearly audible metallic noise. The camel-spiders (Solifugae) – extraordinary relations of the scorpion, with enormous jaws – screech quite loudly when enraged. The sound is produced by rubbing together two rasps on the inner surfaces of the jaws, but whether the threat is directed towards enemies other than rival members of their own species, is uncertain.

Like many other venomous animals, several species of scorpion are capable of producing warning sounds. These are stridulations similar to the hissing produced by rubbing the flat part of a comb across a stiff brush. In some scorpions, the sounds are made by the movement of a 'keyboard' of flattened bristles on the base of the claws against a rasp of finely striated cuticle on the walking legs. In other species the positions of the keyboard and rasp are reversed. The factor decisive in determining which type of mechanism has evolved was probably a difference in behaviour. The response to danger by scorpions of the first group was originally mainly aggressive, and clutching movements of the claws predominated whereas, in the second group, the principal movements were defensive. These differences then conferred selective advantage on the elaboration of devices increasing the noise produced by friction. Other scorpion species stridulate by rubbing their jaws together, and others still by scraping the point of the sting against the first two segments of their tail.

Many poisonous animals, then, including wasps and snakes, make themselves conspicuous by producing sounds which serve as danger signals to meddlesome intruders; in this way the poisonous forms are not destroyed by carnivorous enemies in mistake for harmless and edible species. The threatening buzz of a swarm of angry bees is enough to deter anyone except a beekeeper, while the angry buzzing of wasps and hornets can be equally forbidding. Even small mutillid wasps, or 'velvet ants' as they are sometimes called, produce such a nasty sound that most people and, presumably, many other animals wisely leave them alone. Defensive stridulation is characteristic of mantids – whose spiny forelegs can do considerable damage to small lizards – or grasshoppers that emit poisonous froth, unpalatable moths and other insects. In some instances stridulation may be mimetic or merely bluff, but more often it is aposematic and has a genuine warning function. There is, however, no evidence that scorpions or snakes can perceive the sounds produced by their own stridulating organs. The existence of stridulatory organs implies an auditory sense, not in the performers themselves but in the enemies that might otherwise wound or destroy them before receiving a poisonous sting or bite. In fact, scorpions, like snakes, are deaf to air-borne sounds.

Not until a few years ago was it realized that ground-beetles are able to stridulate warningly when disturbed. A file on the wing is drawn across a scraper or 'plectrum' on the inner surfaces of the wing cases. The mechanism produces soft but clearly audible chirping sounds that are used to scare enemies, such as birds and mice. In a test with sandpipers, which were offered both normal beetles and insects from which the sound-producing apparatus had been removed, it was found that stridulating beetles were released more often and swallowed less spontaneously than silenced ones. Stridulation in these insects is therefore again to be interpreted as a warning signal which may be associated with the defensive secretions.

Sound production is not restricted to the adult stages of insects, but it is otherwise rare except in the pupae of Lepidoptera. Some of these make a knocking or rattling sound by banging their bodies against leaves or other objects. The pupa of one African butterfly belonging to the Lycaenidae, the family that includes *Spalgis*, the blues, coppers and hairstreaks, hammers rapidly on a leaf with its head when alarmed making, it is said, 'quite sufficient noise to frighten away a small predator'. It should be stressed that this, and other similar statements made in the literature, do not constitute acceptable evidence of function. Judgment must be reserved until, from rigorous experiment, proof has been obtained that the sounds really do have a biological significance and are not merely incidental.

Warning and Threat

Stridulating organs are not infrequently to be found on one or more pairs of abdominal segments of a chrysalis. In Lycaenidae and some hawk-moths (Sphingidae) the stridulatory surfaces consist of coarse tubercles. One moth *(Gangara)* has a stridulatory organ on the abdomen. When this is moved, a series of ridges scrape other ridges on the underside of the proboscis, to produce a hissing sound. Movements of pupae that are enclosed in cocoons of stiff, parchment-like silk are frequently audible to humans. Often the chrysalis makes scraping noises with backwardly directed spines, and occasionally there are ridges on the inner wall of the cocoon as well.

Insectivorous bats detect and locate their prey by means of sonar. The ultra-sounds they produce, however, are sometimes detected by moths which take evasive action and make erratic flight movements. In addition, some tiger-moths can themselves produce ultra-sounds which probably serve as a warning that the moth is unpalatable, and may perhaps also interfere with the bat's sonar. Certainly it makes them swerve away. It has been shown experimentally that the moths produce clicks in response to the sonar pulses of a bat over 2 metres distant – far enough for the latter to hear the click and avoid the moth. Some edible moths may even mimic the clicks of distasteful species.

The functions of arthropod sounds are numerous. Usually they are concerned with courtship and aggregation, but stridulation, as a warning, is found in many distasteful and poisonous insects, their larvae and pupae, arachnids, and in crustaceans with powerful claws. At the same time, they may have a social value by warning other individuals of danger. The noise made by the spiny lobster *(Palinurus)*, for instance, causes all other individuals within a range of 1–2 metres to retire into their hiding places.

The menacing hiss of a large snake, like the sound of escaping steam, is more than enough to send cold shivers down the broadest back! So startling is the sound that many non-poisonous snakes bluff their way out of trouble with threatening hisses. In addition to hissing, the African carpet-viper *(Echis)*, whose venom was discussed in the previous chapter, has a dramatic threat display in which the body is thrown into a concertina of tight coils, constantly swirled by throwing waves of movement back down the body. As the coils scrape against one another, the snake's rough scales produce a rasping noise like a handful of

Tooth & Claw

gravel continuously rattled in a small cardboard box. It is similar to the sound of a rattlesnake, whose rattle is composed of a number of interlocking, horny pieces of cast skin which, when rapidly vibrated, produce an angry buzz. It is said that, in the United States, rattlesnakes do not rattle to the same extent that they used to do. Apparently, since the snakes that draw attention to themselves are often killed by Man, there has been severe selection in favour of snakes that do not readily make their characteristic warning sounds!

Many other poisonous and well defended animals advertise their presence, not only by aposematic coloration and warning sounds, but also by scents. The porcupine is strictly nocturnal in habit, but the whiteness of its quills, when erected in a characteristic fan, render the animal conspicuous in the dark. Furthermore, when hard-pressed by an enemy, the porcupine seems to make as much noise as possible, shaking its tail quills – which are specially modified to form a rattle – uttering hoarse and gutteral grunts, and producing a powerful stink. If none of these warning

27. Porcupine.

devices is sufficient to secure the departure of the enemy, the porcupine will launch itself backward with lightning speed and, as we have already seen, deal the enemy a severe blow with the bunch of quills above its tail. Since these are barbed and readily detached, they can cause severe injury to a predator that is unable to remove them. Many lions and tigers become man-eaters because injury to their paws from the quills of a porcupine has made it impossible for them to kill their normal prey. Hissing is employed for purposes of warning by the venomous Gila monster and Mexican bearded lizard, and by various snakes, birds and mammals which hiss before they strike or bite. Even domestic cats do so.

Although venomous animals are usually conspicuous, this is not true of jellyfishes, scorpion-fish, snakes and other animals in which poison is a second line of defence. Quite a number of animals, especially in the tropics where selection is heavier than in other parts of the world, have several lines of defence. *Spalgis* pupae, disguised like the faces of monkeys, resemble bird droppings when seen from afar. Insects and other animals, cryptic from a distance, may display aposematic colours and sounds when discovered. If these fail to drive the enemy away, they will bite or sting, after which, if necessary, they flee. Furthermore, they may be nocturnal anchorites with bodies protected by armour, spines or repugnatorial chemicals.

Warning signals, of course, are not invariably effective in deterring hungry predators – any more than are the defensive weapons they advertise. Even skunks are occasionally attacked, and snakes are forced to bite in self-defence. On balance, however, it usually pays distasteful or venomous animals to display bright colours, scents and sounds which enable potential predators to recognize and thus avoid them. Consequently, it is by no means surprising to find that harmless animals should sometimes cash in on the evil reputations earned by aposematic species and mimic them to deceive their own enemies.

9 Mimicry

While the term 'mimicry' has its origin in the non-technical word meaning 'imitation', it is restricted by zoologists to the sense of close resemblance of one organism to another which, because it is unpalatable and conspicuous or aposematic, is recognized and generally avoided by most predators. While mimicry can only be explained in terms of evolution and the survival of the fittest, mimetic animals were recognized long before 1858, the year in which Charles Darwin and Alfred Russel Wallace proposed the theory of evolution by natural selection. During the eighteenth century taxonomists had called attention to the way in which insects of different orders sometimes resembled one another very closely and, in 1817, W. Kirby and W. Spence noted that certain drone-flies strikingly resembled bumble-bees 'in shape, clothing and color'.

Darwin did not consider the idea of adaptive coloration in the first edition of *The Origin of Species*, published in 1859, and it was actually H. W. Bates, in 1862, who was the first to publish a clear discussion of mimicry – based upon the observations he had made on the butterflies of the Amazon valley. Within a short time, however, mimicry became a pivot in the theory of natural selection. Bates had returned to England with a collection of nearly 15,000 butterflies, no less than 8,000 of which were new to science. When he began to study them he found what he described as 'resemblances in external appearance, shapes, and colours between members of widely distinct families . . .'

The idea of mimicry was based particularly on the study of butterflies of the tropical family Heliconiidae. In South America, Bates had noticed that these common insects were never eaten by birds or lizards, despite their brilliant coloration and slow, conspicuous flight. Surprisingly, other butterflies of a second, quite unrelated family, appeared superficially to be identical with the Heliconiidae. To explain this curious relationship, Bates proposed that the common heliconiids were distasteful to insect-eating vertebrates. Upon trying one, a bird or lizard would find it unpleasant, reject it, and subsequently remember its conspicuous coloration. When it again encountered a similar butterfly it would ignore it. Thus, a palatable butterfly which mimicked it could, by deception, escape being eaten. Bates realized that, to be effective, the distasteful 'model' must be relatively common and its harmless mimic rare. If this were not so, predators would stand a good chance of sampling a mimic rather than its unpalatable model and, consequently, would not learn to avoid the shared aposematic coloration. On the contrary because the mimic was palatable, they might immediately look out for other, similar prey! In fact, the chances of catching a mimic first time and thus of learning the wrong lesson, are relatively rare.

Harmless flies are often cited as mimics of poisonous wasps, bees and ants; bombardier-beetles are mimicked by grasshoppers, ants by spiders, poisonous moths by harmless species and so on. During the century after Bates first outlined the concept of mimicry which now bears his name, over 1,500 scientific and unscientific papers were published, arguing both for and against it; but the theory of batesian mimicry has now been generally accepted. It is a form of primary defence, like aposematism, in that it decreases the probability that any encounter will take place between an animal and a potential predator. Secondary defences are invoked when an aposematic animal has been attacked or threatened. Its mimics, of course, do not possess these secondary defences and must take the chance that their bluff will not be called.

Batesian mimicry is similar to disguise in that it is a means of deceiving enemies. Whereas anachoresis and crypsis can prevent a predator from detecting its prey, however, disguise, aposematism and mimicry merely ensure that the prey is not recognized as something edible. For convenience, the term 'disguise' is usually applied to cases in which an animal looks like something it is not – for instance, a leaf, or the droppings of a bird – whereas 'mimicry' is restricted to the resemblance of one animal to another not greatly dissimilar, but with aposematic attributes.

Successful employment of aposematism and mimicry depends

upon the ability of predators to learn from unpleasant experiences not to attack animals which display warning coloration, sound or scent. Experiments have shown that vertebrates and cephalopod molluscs are able to learn through experience, but there appears to be no evidence that predatory arthropods can learn to avoid aposematic prey. Consequently, it must be assumed that mimicry on land results from predation by reptiles, birds and mammals; and in the sea by predation from cephalopods and fishes.

On account of their poisonous stings, ants, bees and wasps are everywhere respected by predatory enemies, and consequently serve as models for innumerable batesian mimics of one sort or another – from creatures as dissimilar as grasshoppers, mantids, bugs, moths, flies and beetles. One of the main problems which confronts such mimics is to achieve the appearance of having a narrow waist. Spiders already have one constriction between the cephalothorax and abdomen but, in some ant mimics *(Myrmarachne)*, the front part of the body has an additional narrowing which gives the appearance of a head. The likeness of some spider mimics to their models is so good that, when alive, it is almost impossible to tell them apart. The spiders not only resemble ants very closely, and hold up their front legs to look

28. A spider carrying a dead ant, at the same time mimicking an ant doing so.

Mimicry

like antennae, but furthermore they run about in the same jerky way that ants do. The males of one Sudanese species mimic black ants, while the females are mimics of mutillid wasps.

Most ant mimics produce the appearance of a narrow waist by optical illusion rather than by structural modification. White patches on either side of a dark coloured body, leaving only a thin central line, give the appearance of a slender black stalk at the front of the abdomen. Nymphs of a British predatory bug *(Nabis)* produce an ant-like appearance by the same means and various grasshoppers, beetles and flies, in various parts of the world, do likewise. Certain tropical spiders, although not the least like ants in general appearance, nevertheless give an ant-like impression by running about in a jerky manner. At the same time, as mentioned in Chapter 3, some of them carry over their backs the empty dried skeletons of real ants in such a way as to hide their own bodies completely from view.

Ants act as models in particular for many species of jumping-spiders. In one bizarre instance, the spider mimics an insect in reverse. The abdomen of a Bornean spider *(Orsima)* is strongly constricted, with the posterior portion wide and shaped like a head. The spinnerets, the organs with which the spider secretes its silk, are long, giving the illusion of antennae and mouth-parts. The resemblance to an ant is enhanced by the movements and posture of the spider, which often stands with posterior 'head' raised above the horizontal. The antenna-like spinnerets are moved from side to side and, when the abdomen is lowered to attach a silk drag-line, the remaining spinnerets come together and look like the mandibles of an ant contacting the surface. The spinnerets are darkly pigmented, which makes them conspicuous, as are the posterior legs of the spider, while its abdomen is clothed with irridescent blue-green scales. In addition to the advantage conferred by batesian mimicry, it has been suggested that the apparent reversal of the front of the body would allow greater opportunity of escape because a predator's attention would be diverted from the true head and the intended victim would depart in a direction opposite to the one expected. Similar deception has been observed in the posterior antenna-like extensions of the wings of blue butterflies and alligator-bugs.

The only other case of reverse mimicry in spiders has also been reported from Borneo. In this *(Amyciaea)*, dark spots on a light

Tooth & Claw

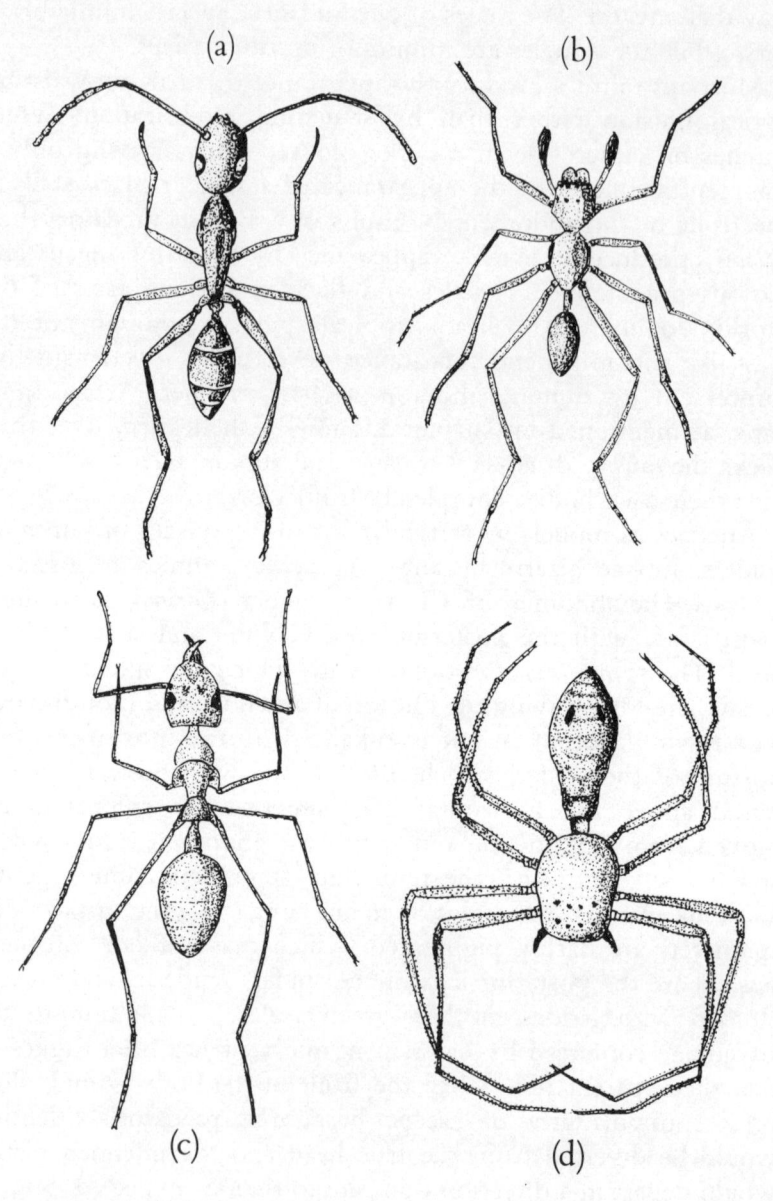

29. Spiders mimicking ants. (a) *Camponotus* and its mimic (b) *Myrmarachne*; (c) *Oecophylla* and its mimic (d) *Amyciaea* (which mimics in reverse). (After L. Berland.)

144

orange abdomen give the illusion of the head of a tree-ant *(Oecophylla)* with its dark compound eyes. In its movements, however, the spider is ant-like in the ordinary fashion. That is, it behaves as though its front end was the ant's head, and does not usually move backwards. Consequently the illusion is only effective when the spider is at rest. Even so, eye spots at the wrong end of the body probably allow greater opportunity of escape if the spider is attacked by an enemy. Many of the animals that mimic ants not only live with their models as uninvited guests, but exploit them – eating their food and even preying on them. Such mimicry was extensively discussed by Eric Wasmann and has been named wasmannian mimicry after him. It will be discussed further in Chapter 14.

Mimicry is probably more widespread among insects than it is in any other group of animals. It is comparatively rare in terrestrial vertebrates, apart from a few fishes, snakes and birds. The most conspicuous case of batesian mimicry in mammals has been cited from Borneo. Here, five species of tree-shrews, whose flesh has a repulsive taste, are mimicked by five species of palatable squirrels which resemble their models so closely that skins from the two groups cannot be separated, and the mimetic pairs can be distinguished only by means of their skulls.

The suggestion, recently made, that the small and inoffensive aardwolf *(Proteles)* may have evolved its close visual resemblance to the larger striped hyena for protection against visual predators, mainly leopards, is therefore of interest to evolutionary biologists. The two species resemble each other in general habits, some aspects of colour and pattern, and in defensive behaviour. Like the hyena, the aardwolf has a sloping back, pointed ears, and a well developed, erectile mane alone the dorsal spine from neck to tail. This mane is composed of stiff hairs, some 20 cm in length and, when erected as a threat display, makes the aardwolf appear considerably larger than it actually is. Leopards prey on jackals, but not on hyenas or aardwolves. This is understandable in the case of hyenas which are equipped with formidable teeth and strong jaws; but aardwolves have greatly reduced dentition – a specialization for eating termites – and are consequently more or less incapable of active defence against a large predator. There is therefore circumstantial evidence for the hypothesis that the aardwolf is a batesian mimic of the hyena but, before a definite

Tooth & Claw

conclusion can be drawn, further information is required about the spatial and temporal overlaps of the two species, their relative palatability compared with that of the jackal, and the reaction of leopards to the erection of their manes. As in so many instances of apparent mimicry, probability is high but proof is lacking.

Two cases of bee mimicry have, however, been investigated experimentally in central Florida and clearly established. One concerns a bumble-bee *(Bombus)* which is mimicked by a robber-fly *(Mallophora)*. The two insects are seen together quite

30. *Above,* aardwolf and *below,* hyena, showing resemblance. (Drawn from photographs. Not to scale.)

frequently in fields where leguminous plants are flowering. The robber-fly has a black and light brown pattern, a plump fuzzy body like the bee, and hairy legs, on the third pair of which it even has two patches of lighter hairs that simulate the pollen baskets of the bumble-bee. The second model is the honey-bee *(Apis)* whose mimic is a drone-fly *(Eristalis)*. The drone-fly has a narrow black line along the centre of its back which creates the impression of the honey-bee's narrow waist. It also has black and yellow rings round its abdomen, like those of its model and, when feeding on flowers alongside honey-bees, it buzzes like the bees do.

Experiments were carried out by L. P. and J. V. Z. Brower in which toads were used as predators, and offered bumble-bees. After being stung, the toads learned to reject bumble-bees, and also robber-fly mimics, on sight. Toads that had never experinced a bumble-bee, however, readily ate robber-flies, and also took bees from which the sting had been removed. A second series of experiments showed that live honey-bees also stung the toads, although not so severely as the larger bumble-bees, and were subsequently rejected. Nevertheless, after experiencing a honey-bee sting, drone-flies were rejected although, if offered first, they were readily eaten – showing that they were perfectly palatable and, therefore, true batesian mimics.

Predatory vertebrates, including toads, lizards, birds and mammals, do not find their food by continuous random searching. Instead, they form what is known as the 'searching image of a specific prey' in their hunting behaviour. This means that they acquire a general idea of the appearance of edible prey, which they look for – even to the extent of ignoring other, equally palatable animals. How often, in a supermarket, does one overlook the very item for which one is seeking because the label on the tin has been altered, or because a different brand is stocked! The more often predators see their prey, the more the searching image is reinforced. In a broad sense, animals with cryptic coloration derive protection from mimicking the non-living substrate, flecks of sunlight or living plant materials in their environment. This tends to foil their predators' searching images by invoking an opposite kind of learning, known as 'habituation'. Habituation means learning *not* to respond to a stimulus that has no significance in the life of the animal. Animals become tame by habituation: when they find they are treated kindly, they cease to react by flight to the presence of a human being. Now, if a bird predator were to discover a tasty stick-insect, for example, concealed in the branches of a tree, it would probably search for more of them; but, because of the insect's close resemblance to the abundant twigs, the bird would have difficulty in finding them. Everytime it mistook a twig for a stick-insect, it would neither be punished nor rewarded for its mistake. Consequently, it would become habituated to the searching image of a stick-insect and would begin to look for something else instead.

Predators learn to avoid unpalatable prey, with aposematic coloration, through another type of learning, known as 'conditioning'. In this, they come to associate the bright colour pattern and conspicuous behaviour of the prey with its unpalatability. After one or two unpleasant experiences, a predator soon learns to avoid animals which are distasteful, poisonous or have other formidable defences: and, as a result of conditioning, they also avoid their mimics. The colour, pattern and display behaviour of saw-fly caterpillars are all eminently adapted to promote conditioned visual rejection by bird predators. Similarly, bats, which hunt by echo-location rather than by sight, learn to avoid unpalatable tiger-moths, as we have seen. These exude nauseous smelling fluids when attacked, but make warning sounds so that bats leave them severely alone.

A noxious species only benefits from warning coloration because some of its members are sacrificed in teaching would-be predators to avoid it. Therefore, if one or more aposematic species mimic one another, the numerical losses incurred in teaching enemies not to attack them are shared among them and proportionately reduced. The mimicry of one distasteful model by another distasteful species is known as müllerian mimicry, after F. Müller who pointed out, in 1878, that the number of individuals sacrificed to each predator could be spread over a number of species to the mutual benefit of each. For instance, many species of wasps and bees mimic one another. So do inedible moths of quite different families. Water-mites have integumental glands which render them distasteful to sticklebacks and other small predatory fishes: the conspicuous colour red is common to many, quite unrelated, species.

Müllerian mimicry, in which one noxious animal mimics

Mimicry

another to the benefit of each, is not confined to aposematic coloration and visible signals. The hisses of different kinds of poisonous snakes appear to be very much alike, as do the buzzing sounds made by bees and wasps. In rather few cases, however, has the perfection of the mimicry of the warning sounds actually been analysed. An example of such analysis is provided by a study of burying-beetles whose stridulation is a well-known phenomenon. Two transversely striated files set at an angle, one on each side of the dorsal surface of the fifth abdominal segment, are rubbed against the tips of the wing cases to produce a rasping noise. Analyses of recordings of this, and of the buzzing of bumble-bees, show that the sounds have the same general form. Cathode ray oscillograms and sound spectrographs of the two are remarkably similar.

The stridulation of burying-beetles, however, is only part of an elaborate pattern of behaviour reminiscent of that exhibited by bumble-bees when disturbed in a semi-torpid state. When molested, a burying-beetle will often turn on its back and begin to stridulate. In this position, the first pair of legs is held upright at an angle to the body, and the second and third pairs are pushed out laterally like oars in a manner characteristic of the movements of the legs made by a bumble-bee when disturbed while resting. At the same time the beetle works the tip of its abdomen vigorously in and out, thereby rubbing the files against the serrations and producing a series of short bursts of stridulation. These movements not only resemble the stinging movements of a semi-torpid bee, but the noise emitted is a fair imitation of the latter's short bursts of peevish buzzing. Furthermore, during the display, the tip of the beetle's abdomen becomes covered with a white or brownish froth which, together with the light-coloured belt of hairs on the abdomen, and the reddish-brown edge of the wing case seen from below, give a general, if rather crude, resemblance to the colour scheme of a bumble-bee. Although burying-beetles themselves secrete a potent repugnatorial liquid, and have the disconcerting habit of squirting it at the source of any disturbance, they probably gain added protection from mimicry of bumble-bees, which have venomous stings.

In all parts of the world, the larger species of parasitic Hymenoptera – ichneumons and their allies – exhibit striking colour patterns. They mimic the local wasps, lycid beetles and other protected forms. The fauna of tropical regions tends to be more brightly coloured than that of temperate areas. The Oriental region is dominated by a dazzling black-and-white pattern which is amazingly uniform in thousands of species. Several mainly yellow patterns are found in day-active species, the commonest of which is completely yellow except for the eyes and the tips of the antennae. Other yellow patterns have black spots or variously coloured wings, although the latter are relatively scarce. In Africa, on the other hand, species with coloured wings are abundant. One pattern consists of black abdomen, thorax, legs and wings, with a contrasting yellow head and anterior legs: another is similar, but the thorax is red and the head black. There are more variants of the yellow or mostly yellow pattern in Africa than elsewhere in the tropics, but the tropical parts of South and Central America contain the greatest variety of colour patterns in the world. Most of the insects with these patterns have clear wings, but one especially common one in the northern parts of South America has yellow wings with two transverse brown bands. The body is yellow while the head, and often the tip of the abdomen, are black. Müllerian mimicry is the probable explanation of these groups of colour patterns because there are no dominant 'model' species involved and because, although the sting of an ichneumon is far less effective than that of a social wasp or bee, it does give some protection.

Other well-known examples of mimicry come from the American tropics where there are four or five major mimicry complexes involving large numbers of species. One extensive group of beetles, from diverse evolutionary lines – many of them weevils – has a striking colour pattern consisting of a red anterior area, a black or variegated mid-section and a larger grey or yellowish posterior area. It seems probable that the pattern is meant to mimic medium-sized to large flies, and the variety of patterns among the beetles implies that a number of fly models are involved. The behavioural characteristics shared by both the beetles and the flies is their tendency to perch on relatively isolated and exposed tree boles where they can easily be seen and readily mistaken for one another. The question arises however: what is the advantage of looking like a fly? Flies are not known to be generally distasteful, and there is no evidence that the beetles are. The suggestion has therefore been made that the basis of the mimicry rests not on chemical distastefulness but on the speed and manoeuvrability of the flies. In general, birds and other predators, which hunt by sight, are seldom able to capture flies of the size and type mimicked by the beetles. Indeed, they probably do not find it worthwhile even to try. Since most of the beetles are also very quick and elusive, it seems probable that the system is functionally müllerian, but that both flies and beetles are avoided by birds not because they are chemically unpalatable but, rather, because it is not worth the effort of trying to catch them! Similarly, in the Namib desert, conspicuous orange wing cases advertise a complex of unusually speedy scarab-beetles.

Bird coloration has likewise evolved almost entirely in response to predator-based selective pressures. Not only are aposematic birds often distasteful, as we have already seen, but bright colours may also indicate to potential enemies that a bird is unprofitable prey – because it is wary, mobile, and difficult to catch. Predators soon learn that it is not worth while trying to catch conspicuous species and therefore tend to leave them alone. In contrast, birds not so good at escaping tend to be immobile and inconspicuous. (Although plumage and colour are involved in species or sex recognition, they are not now believed to have evolved as a result of sexual selection, as Darwin thought.) If bright colours do, indeed, signify unprofitable prey, we might expect them to be mimicked by rather immobile species. But, since aposematic coloration is, in the case of birds, always associated with mobility, it may not be possible for inactive species to mimic unprofitable prey convincingly.

Certain aquatic beetle pupae mimic their own larvae. Because of their shape, the latter are pressed by water currents firmly on to the substrate so that they are not easily eaten by fishes. The pupa, which mimics them, is glued firmly to the substrate. The selective value of their resemblance to larvae appears to lie in the fact that fishes have only to learn one form and pattern instead of two, and so make fewer attempts to pry off the pupae than they otherwise would.

As pointed out earlier, the palatability of certain insect prey to their predators varies according to the food plant eaten by the prey during their development. In the most thoroughly studied example – that of the monarch butterfly – adults that lay their eggs on poisonous milkweed plants produce offspring which are severely emetic when eaten by birds, whereas larvae feeding on non-poisonous milkweeds become highly palatable butterflies. Since predators are unable to distinguish visually between palatable and unpalatable individuals of the same species, an unpleasant experience with an unpalatable butterfly will teach the predator to reject all monarch butterflies. Consequently, an automimetic advantage is an inevitable consequence of the evolution of both palatable and distasteful individuals in the same species. As in the case of batesian mimicry, the mimetic advantage depends upon the proportion of unpalatable to palatable prey, the memory span of the predators, and so on.

Amongst snakes, some harmless species mimic the pattern and warning displays of poisonous ones. In the New World venomous coral-snakes are mimicked by various harmless, or only mildly poisonous 'false' coral-snakes. The colour patterns of models and mimics vary considerably in the arrangement of the bands and in their width, but with a sinuously moving snake it is difficult to recognize a particular species with certainty, as movement blurs the colours. In the past, herpetologists have sometimes rejected the idea of mimicry in the case of coral-snakes because, if a predator were to attack so deadly an adversary, it would almost certainly be killed and, therefore, could never learn to avoid coral-snakes of any kind in future. Secondly, in North America, some of the harmless 'false' coral-snakes occur further north than their models. The question then is, if the coloration of coral-snakes is related neither to kinship nor to poisonousness, what does it depend upon? The solution to this problem is believed to be as follows: the harmless species are typical batesian mimics of the moderately poisonous forms, while the deadly elapids are 'mertensian' mimics of the same mildly venomous species.

Mertensian mimicry (named after R. Mertens who carried out much research on coral-snakes) is said to occur when a deadly species mimics a less dangerous model. If a predator attacked a coral-snake it would probably die, as mentioned above; but possibly not before it had wounded the snake, which might therefore be sacrificed to no purpose. Learning would not have taken place and other coral-snakes would not be protected by their coloration. In fact, venomous coral-snakes are far less numerous than the mildly poisonous, 'false' coral-snakes they mimic. Hence predators are most likely to encounter one of the very aggressive but only slightly poisonous species of snake which bite without killing and therefore cause all coral-snakes to be avoided in future encounters. On the other hand it seems unlikely that any snake could be so poisonous that all animals bitten by it are killed. Furthermore, many predators have both natural and acquired immunity against snake and other venoms.

In areas where their models are absent, the pattern of the harmless 'false' coral-snakes is not in conspicuous rings of black, red and yellow but in blotches and saddle-shaped marks which probably have a disruptive function and break up the outline of the snake. The one deadly species of coral-snake on the island of Tobago also lacks annular markings. Presumably selection must favour aposematic coloration where coral-snakes and their mimics are common and frequently encountered by enemies but, where they are rare, it is of greater protective value to be camouflaged.

In the Old World, most of the poisonous snakes are cryptic. Although they may be poisonous, they are not invulnerable to attack, so it is probably of greater protective value for snakes to be inconspicuous than to have aposematic coloration unless, as in the case of coral-snakes and 'false' coral-snakes, they are extremely numerous. Once a snake has been discovered, however, it is still advantageous to it if the enemy can be frightened away before it has launched an attack. At this point, secondary defences are invoked, and hissing and threat displays are initiated.

It seems reasonable to ask why it should be better for an animal with aposematic coloration *not* to be lethal. It is obvious that learning to avoid an unpalatable species is dependent upon surviving the experience of having attacked one in the first place, but this does not explain why an animal would not be better protected if it were to eliminate its enemies altogether. The answer lies in the fact that most predatory vertebrates are territorial and drive rivals away from their feeding places. If each predator were killed when it attacked or ate poisonous prey, its former territory would be re-occupied by another inexperienced predator. On the other hand, if the poisons of the prey are sufficiently unpleasant to teach the predator a lesson which it does not forget, fewer prey need to be sacrificed to maintain an educated population of predators. Furthermore, most young mammals and birds learn much from their parents, and it is not inconceivable that they should learn to avoid certain aposematic characters without needing to experience the defences these advertise. Not only do both predator and prey species benefit if the poison of the latter is not lethal, but the community as a whole gains stability.

Few biological concepts have aroused more controversy than those of adaptive coloration and mimicry, so it will not surprise the reader to learn that the concept of mertensian mimicry has not only failed to win universal acclaim but, rather, seems to have engendered even more attempted explanations of the colours of coral-snakes, true and false. In the United States there is one highly venomous coral-snake and two species of false coral-snake which are non-poisonous: there are no mildly poisonous forms. One could, of course, assume that a mildly venomous species, perhaps a progenitor of the highly venomous coral-snake, at one time inhabited the area and served as a model and then became extinct. But there is no evidence to support such an assumption, nor to indicate that any of the species now found in the United States has migrated from elsewhere. The suggestion has therefore been made that predators may have an innate aversion to red-yellow-black ringed patterns, although how this has been acquired is not explained. An alternative suggestion – that similar colour patterns have arisen independently of natural selection in unrelated species occupying similar habitats – is open to the objection that no selective agency is postulated. In this context it is worth remembering that coral-snakes of all kinds are nocturnal and are usually concealed from day-active predators with colour vision. They would normally only be seen by nocturnal animals which lack colour vision and would therefore be unable to appreciate the details of their bright colouring. We may be certain that further hypotheses will appear regularly for many years to come!

In both batesian and müllerian mimicry, the mimic does not require anything from any other animal; it merely wants to be left alone. Just as angler-fish and the matamata use disguise to further aggressive designs against their prey so, in aggressive mimicry, a predatory mimic resembles a harmless model. An example is afforded by the rare American zone-tailed hawk which glides in the company of vultures. Unlike buzzards, to which it is related, the zone-tailed hawk is nearly black, its wings are long and narrow like those of a vulture, it does not use a look-out perch, and does not hover. Instead, it glides like the vultures it mimics. Now, vultures present no threat to small animals, which ignore them. The zone-tailed hawk takes advantage of this, diving suddenly from a group of vultures to take its prey by surprise.

Another example of aggressive mimicry is afforded by the recent observation that females of a number of firefly species *(Photuris)* prey on male fireflies of other species. Whilst all fireflies attract members of the opposite sex by specific flashing light signals, *Photuris* females attract males of different species by adjusting their own responses to those of the prey, thereby mimicking the flash signals of the prey's own females. The robber-flies, which L. P. and J. V. Z. Brower showed experimentally to be batesian mimics of bumble-bees, also use their resemblance for aggessive purposes. Bumble-bees are their favourite food, and they are more successful in catching these than are other species of robber-flies which do not resemble bees so closely.

The concept of agressive mimicry is employed every time a spy disguises himself as a harmless citizen of the opposing country. Many a prisoner has made his escape across enemy territory wearing false uniform. In October 1906, a cobbler named Wilhelm Voigt became famous as the result of a successful stunt incorporating the principles of aggressive mimicry. Donning a stolen army captain's uniform, Voigt stopped a number of soldiers in the street, led them to the town hall of Berlin – Köpenick, arrested the mayor and confiscated the contents of the city treasury. Beside this feat, the escapade of the Cambridge under-graduates who dressed up as workmen and dug a great hole in Piccadilly pales into insignificance. Both, however, depend upon the fact that human beings tend to base their judgments upon relatively few visual signs; and predators do the same.

Mimicry is a topic of such general interest and biological importance that whole books have been devoted to arguments for and against its very existence. Thanks, however, to much painstaking research and the assembly of data by E. B. Poulton and, more recently, by G. D. H. Carpenter, H. B. Cott, L. P. and J. V. Z. Brower, W. Wickler, M. Edmunds and others, the concept of mimicry is now generally accepted, and the main controversy seems to have shifted towards consideration of its genetical applications. In order to restrict the topic to the confines of a single chapter, I have limited somewhat the number of classical examples cited, at the same time concentrating on newer observations and ideas.

10 Bluff, Death Feigning and Deflection of Attack

The Greek god Proteus was said to escape from his enemies by changing his form. Many kinds of animals, too, are able to evade predators by making dramatic changes in their appearance. The alligator-bug, for instance, is cryptic when seen from afar but, close up, the patterns of the hollow extension of its body give a remarkable illusion of an alligator's head. The insect pupae that look like bird droppings from afar and miniature monkey faces close up, provide another example. Others include the caterpillar of a large tropical American sphinx-moth *(Leucorhampha)* which lives upside-down among wild vines and creepers along the edge of the forest. It is extremely difficult to make out from a distance but, if the vine is shaken, the larva lets go with all except its hind legs, swells up in front to expose previously hidden but most realistic dummy eyes, and immediately begins to sway in a sinuous, snake-like manner. When touched, it lashes its head back and forth like a viper striking. In Trinidad, yet another sphingid caterpillar *(Pholus)* also looks like a snake. The resemblance even runs to a forked tongue, simulated by lighter coloration along the head sutures. If all this is not enough to deter, the caterpillar also has a structure on its posterior end which resembles an enormous winking eye!

Such unexpected displays, sometimes called 'protean' – after the versatile god – are difficult to classify. Is a leaf-insect, invisible against a background of green foliage, merely cryptic, or does it show protective resemblance? Different principles are involved: in crypsis, the prey is not detected while, in protective resemblance, it may be seen but is not recognized for what it is. Perhaps the leaf-insect's first line of defence is crypsis – it is not noticed from a distance – and protective resemblance is only involved at close range, as a second line of defence. Spies are often furnished with two false indentities. If they are picked up and the first of these is penetrated, their interrogators may well accept the second pretence and never discover the real truth. The alligator-bug and the insect pupae that look like monkey faces were discussed as examples of disguise. They could, perhaps, have been regarded equally well as examples of batesian mimicry. Or, is a third factor involved – 'deimatic' (from the Greek word meaning 'to frighten') or startling behaviour? When a grass-snake hisses, is it mimicking an adder or is the sudden hiss alone sufficient to deter an enemy without any element of mimicry being involved? Nature is subtle and complex: only Man can be analytical. No doubt many animals employ various combinations of several different principles in their need to avert enemy attack.

The advantage of deimatic or frightening coloration lies in the fact that it exploits still other aspects of the behaviour of vertebrate predators that have not yet been mentioned. When a hunting animal is suddenly confronted with something it has never before seen, it will often respond by running away or, at the very least, its motivation for attacking may be greatly lowered. The extent to which this effect is induced is apparently related to the novelty and complexity of the stimulus, which both surprises and frightens the predator. Experimental studies have suggested that the smallest predatory vertebrates, such as insectivorous birds, are the most easily affected in this way. Perhaps because they, themselves, serve as prey for larger predators, they are always cautious and wary. If, when searching for caterpillars, they were to encounter a snake-like species, they would probably react as though the caterpillar were one of their own predators and fly without attempting to call its bluff. A different type of deimatic response is seen in orb-web spiders which rapidly vibrate themselves and their webs when they are disturbed. The animal becomes blurred in appearance, possibly with an apparent increase in size, and it must be difficult for a predator to attack such a target successfully.

Deimatic anti-predator displays – sometimes bluff, sometimes genuine warning of noxious qualities – are exhibited by many kinds of insects, including stick- and leaf-insects, mantids and grasshoppers, whose disguise has been uncovered by an enemy. One African mantid *(Idolium)* has forelegs swollen in size and

Bluff, Death Feigning and Deflection of Attack

conspicuously coloured. When one of these insects is approached by a monkey it rears up and displays its formidable limbs. The monkey backs away! Many moths which characteristically rest on vegetation exhibit deimatic displays if they are disturbed. In the case of hawk-moths, which are quite palatable, these displays are bluff. They deter the predator and allow time for the moth to warm itself by rapid wing vibration before it escapes. Other moths and pupae expose bright colours, or produce squeaks and other sounds which intimidate potential predators, causing them to refrain from attacking.

Increase in size, real or apparent, is a common feature of deimatic displays and bluff. When disturbed, the puffer-fish inflates its body with air until it resembles a balloon, and then floats conspicuously upside-down. Although these fishes are poisonous, chameleons and iguanid lizards, which also inflate their bodies to intimidate enemies, are not distasteful – their displays are mere bluff. Many toads and frogs, some of which are toxic, others not, blow themselves up in defence. This is an instinctive reaction of the common European frog when confronted by a grass-snake and it also makes the frog much more difficult to swallow than it would normally be. Inflation is automatically triggered by the sign-stimulus of any long, snake-like object; so the effect of a garden hose on a frog or a toad can easily be imagined! Under provocation, the South American horned frog or escuerzo *(Ceratophrys)* will distend its body to a tremendous size, at the same time uttering broken cries of remonstrance, while a giant toad from the same continent inflates its lungs until it becomes quite discoidal in form.

Not only does the chameleon when approached by a predator or rival inflate its lungs until it almost bursts, but it also turns its body so that the broadest part is exposed. When displaying before an enemy which is looking down on it from above, a chameleon will twist into an almost horizontal position in which, with lungs and throat fully inflated, it presents an abnormally formidable appearance to the foe. When a toad puffs itself up, it flattens out, the body distended laterally. This warning display results in a most comical attitude being taken up when a toad is approached from the side: so that the widest part of its body should be tilted towards the enemy, the near side is pressed against the ground while the further limbs are fully extended.

Tooth & Claw

Inflation of the body is not confined to harmless amphibians and reptiles, for venomous snakes also increase their apparent size during threat displays. The African boomslang *(Dispholidus)*, for example, inflates both its trachea and its lung until it looks like an enormous sausage! Cobras likewise create the impression of increased size when they spread their hoods which have conspicuous markings. If they can scare an enemy in this way, cobras do not need to waste venom on it. Nor do animals rely only upon inflating their lungs to make themselves bigger. Apparent increase in bulk may be achieved by the erection of spines, scales, feathers, hair or quills. The spirit of Hamlet's father could, we are told, have unfolded a tale whose lightest word would have made his indecisive son's hairs to stand on end, 'like quills upon the fretful porpentine'.

31. Indian cobra with hood expanded. (Drawn from photographs.)

Bluff, Death Feigning and Deflection of Attack

Certain poisonous animals have conspicuous colours on their ventral surfaces which are displayed only when all else has failed to deter attack. This is true not only of the aposematic coloration of many venomous snakes, but also of the fire-bellied toad which turns on its back in moments of danger. Although they are really quite harmless, sphinx-moth caterpillars rear up suddenly to expose their terrifying ventral eye-spots. Various cryptic moths, grasshoppers and mantids likewise display conspicuously marked wings or flattened limbs when threatened. In the case of noxious species, deimatic behaviour has a genuine warning function but, in the harmless forms, it is pure bluff. Even for poisonous species it is preferable not to have to use in self-defence venom that would be better employed for the capture of prey while, in the case of the fire-bellied toad and other animals whose poison can only be experienced after a predator has attacked, the advantage of not being bitten is obvious.

Surprisingly, some of the mammals that are most effectively protected against predation have evolved the most elaborate deimatic displays. In such cases, of course, bluff is not invoked since the animals concerned are genuinely formidable and unpleasant in one way or another. When a skunk, for instance, is threatened by a predator, it erects its tail and stamps its forefeet on the ground; while the spotted skunk, as already mentioned, rears up on its forepaws. The zorilla and African polecat show similar responses to enemies. Porcupines erect their quills and rattle them before attacking, while dogs, wolves and other carnivores raise the hair on their backs. It is interesting that the aardwolf *(Proteles)* also erects its hair in the usual way when threatened, thereby possibly mimicking a hyena but, unlike typical carnivores, keeps its mouth closed. This, no doubt, is because it has only rudimentary teeth and to expose them would demonstrate to a predator that it had no real weapons of defence.

As with visual colours and displays, auditory signals may sometimes represent real warning: sometimes they are mimetic, sometimes deimatic. Rattlesnakes and porcupines both make rattling sounds which are genuine warnings; the viper and the grass-snake both hiss, although only the former is venomous. The hissing of the grass-snake may perhaps be batesian mimicry although, like the hiss of a chameleon or of a monitor lizard, it could well be mere bluff. The warning displays of many birds, such as snipe and parrots, are auditory as well as visual. Hole-nesting species such as tits and the wryneck, make a snake-like hiss if a predator appears at the nest entrance. In nocturnal marsupials (Dasyuridae), threatening is normally accompanied by panting or hissing sounds at low intensities of stimulation, rising, at high intensity, to a shrill squeak which, in smaller species, sounds almost like an insect's stridulation. Rodents chatter and gnash their strong incisor teeth in threat, as do pigs, camels and dromedaries.

Deimatic noises, like those uttered by aposematic animals and their mimics, are seldom very loud. The reason for this can be explained in terms of function. Unlike the signals by which rival males proclaim their territories and summon their mates – which are emitted loudly and repeatedly – warning sounds are only used at close range and when an attack is imminent. Their very softness, and unexpectedness, makes them sinister in our ears, and, to judge by the reactions elicited, we may be sure that other animals feel the same.

Among carnivores, threat is normally accompanied by growling, snarling or screeching sounds. In some species, the opening of the mouth is modified in such a way as to accentuate display of the canine teeth which constitute the main weapons of offence and defence. When a dog snarls, the lip is drawn up so that it bares the canine and exposes its entire length. Horses, asses, zebras and some deer also display their canine teeth when threatened, but it is in the hippopotamus that display of the teeth is most impressive. The mouth is opened widely and the head moved round in a circular path, as though the animal were yawning.

Camels and dromedaries threaten enemies, as well as rivals, with teeth gnashing, heads held high and necks displayed to demonstrate their size and strength. This display is basically a threat which has been ritualized as a form of intimidation directed towards sexual rivals. When they move in to attack, however, camels lower their heads. The vicuña enhances its apparent height by standing on any convenient mound or tussock of grass when displaying its head and neck to a rival. Antelope and deer threaten rivals by displaying horns and antlers respectively, but here we are crossing the boundary between defence against predators and territorial displays, for many of

Bluff, Death Feigning and Deflection of Attack

these animals use their horns quite differently according to circumstances. The male giraffe fights ritual battles with his neck and tiny horns, but uses his most formidable hooves to defend himself against a lion.

Deimatic behaviour inhibits attack by bluffing the enemy into thinking that its intended prey may be more formidable than it really is. Another method of inhibiting an attack is by feigning death so that the prey fails to release a killing response in its predator. The best-known example of this is seen in American opossums *(Didelphis)* from which we get the expression 'to play possum'. Many insects of various kinds become inert when they are attacked, either with their legs extended, or withdrawn close

32. American opossum feigning death. (After F. Bourlière.)

to the body so that they cannot be bitten off. Death feigning is advantageous to an animal that cannot escape because many predators, including cats, lizards and mantids, strike only at prey which shows signs of life. When cats play with an injured mouse, it is easy to see how the mouse might do better by remaining motionless, as though dead, than by trying to run away, only to be recaptured again and again by the cat.

Some snakes roll themselves into tight, motionless balls which may help to protect their vulnerable heads when they are attacked. The grass-snake does this when its repulsive smell has failed to deter an aggressor. Another advantage of death feigning is that the predator may relax its attention and thereby allow the prey to recover and escape unexpectedly. Small birds often rest motionless if captured and held loosely in the hand. Then they suddenly revive and fly off without warning. The corncrake feigns death, but immediately runs away when the attacker ceases to pay close attention to it, as is sometimes the case when an attack has not been motivated by hunger. Death feigning is also exhibited by rails and tinamous among birds, and by foxes, wolves, dingoes and ground-squirrels.

In flight, the Egyptian and ruddy shelduck are magnificent birds, with broad wing pinions of black, white and bronze-green. When these birds alight and close their wings, however, their appearance changes as rapidly as that of Cinderella when the clock struck midnight. This is a clever trick by which the vision of any bird of prey can be baulked. As the black-and-white wing colours suddenly disappear, the sandy-coloured body of the Egyptian goose or shelduck becomes practically invisible. In the same way, the scarlet or blue underwings of desert grasshoppers are suddenly eclipsed when the insects settle: and to rediscover them is like trying to find a particular pebble on the shore!

Colours which momentarily flash into conspicuousness when an animal moves, but are normally concealed, help to deflect the attacks of enemies by misrepresenting the whereabouts of the prey. When a red-underwing moth *(Catocala)* flies, its hindwings, red with bold black markings, are extremely conspicuous. As the insect settles, however, the cryptic forewings are folded back so as to overlie completely the hindwings. Colour schemes of this type are fairly common in butterflies and moths, as well as in grasshoppers, leaf-insects (Phasmidae), cicadas, and other plant-sucking bugs. The sudden disappearance of colour, combined with the equally sudden suspension of movement, tends to mislead the eye and render the animal's exact whereabouts very difficult to detect after it has alighted. An analogous situation is found among flying lizards *(Draco)*. In several oriental species, the flight membrane, supported by a series of modified ribs, by which the lizard glides, bears conspicuous black, orange or yellow 'flash' coloration. It is only when expanded in use that the colours are momentarily revealed, while the lizards glide from branch to branch. Normally the membranes are folded at the sides of the body where they are quite inconspicuous.

Readers will no doubt remember how the wife of Darzee the

Bluff, Death Feigning and Deflection of Attack

tailor-bird fluttered desperately on the ground, pretending her wing was broken, in order to lure Nagaina from the rubbish heap so that Rikki-Tikki-Tavi could destroy the cobra's eggs. Rudyard Kipling obviously knew how birds will feign injury and hence invite pursuit in order to lead predators away from their nests. When frightened from its egg, the South American dusky nightjar *(Caprimulgus)* limps away and crouches in a conveniently exposed spot so that the attention of the hunter is attracted to the bird herself and away from the egg. In cases such as this, a cryptic animal makes itself conspicuous, thus attracting the attention of a predator away from its eggs or young. Gilbert White wrote that 'a partridge will tumble along before a sportsman in order to draw away the dogs from her helpless covey'. Again, the actions of the parent simulate those of a wounded bird, and hence encourage pursuit by the predator.

Injury feigning is chiefly to be met with the birds that nest on the ground. It also occurs in warblers which usually make their nests in bushes at no great height, among the pigeons – a family of which certain members still breed on or near the ground, although others have taken to arboreal nesting – and in the owls, of which the same may be said. The species in which this behaviour takes place with any regularity are few. The best example of all, perhaps, is provided by Arctic skuas – which become extremely demonstrative when their nests are approached, even before they contain eggs. They will throw themselves on the ground and flutter feebly about as if with an injured wing, uttering plaintive cries and seeming to invite pursuit. Oddly enough, when you reach the nest and stand by it, the actions of the birds are not raised to their highest pitch of intensity. Instead, they stop and watch intently as though they realize that it is useless to pretend any longer. With pratincoles, waders, ducks and some game birds, such as the partridge, it is also quite a regular procedure. In most species, on the other hand, it is remarkable how seldom injury feigning can be evoked by Man. This may be because Man does not normally inspire fear in birds, as predators do, except in places where there is regular shooting. In most instances, harmless intruders – such as sheep and cows – are driven off by curlews and plovers which fly into their faces. The furious stoop of the great skua at one's head and the swooping attack of the brown owl are merely the reaction to a large creature which must be intimidated or driven to one side: Man is clearly not always recognized as a potential predator on the eggs.

Of all parts of the body, the eye is probably the most difficult to conceal because of its regular shape and characteristic coloration. The eyes of many cryptic animals, as we have noted, are masked or hidden by a variety of devices which are often extraordinarily effective. It is, therefore, not surprising to find that false or imitation eyes have been evolved whose function is to deflect the attack of a predator to a part of its prey that is less vulnerable than the head. Dummy eye-spots are found in a wide range of animals. They are to be seen on the wings of birds, butterflies, mantids and other insects, and on the bodies of spiders, insects, fishes, frogs and reptiles. Their effectiveness is demonstrated by the numbers of butterflies and moths to be found with characteristic V-shaped imprints, caused by the beaks of birds, at or just beside the eye-spots. On occasions, the directive function of the dummy eye-spot is reinforced by the entire shape and pattern of the wings. In one Central American blue butterfly *(Thecla)*, a series of broad black-and-white bands cross the forewing and continue without interruption on the hindwing to converge towards a false head with imitation eyes and even a dummy antenna at the hind extremity of the wing. The same effect is produced in related South American species of the genus *Pseudolycaena* but, in these, the wings are spotted and not striped.

Despite their lethal venom, kraits are not aggressive. When threatened, they crawl slowly away keeping the head close to the ground while the tail is reared upward, like that of an amphisbaenian, but bent slightly towards the object that threatens. As it is blunt, curved at the correct angle and slowly waved, the tail looks like an awesome head awaiting a chance to strike. The effect is enhanced by the fact that the imitation 'throat' of the simulated head is bright red. If this seems bizarre, it will be even more surprising to learn that the dummy red throat on the krait's tail serves as a model, and is mimicked by red colours on the genuine throats of non-poisonous snakes which inhabit the same regions of the world that kraits do. Likewise, some large tropical scolopendras have brightly coloured front and rear ends which seem to be aposematic and mimic one another. These centipedes not only inflict poisonous bites, but can also pinch with their last pair of legs.

Bluff, Death Feigning and Deflection of Attack

In most instances of disguise and protective resemblance, the animal simulates the appearance of some object that is normally ignored or avoided by its predators – for instance, a butterfly may look like a dead leaf or a caterpillar like a snake. Sometimes, however, the animal does not change its appearance. Instead, it alters its surroundings so that it blends into them, or else makes a number of dummies of itself that increase its chances of escaping attack.

Many fine examples of this are provided by orb-web spiders. These are large and usually soft-bodied animals which provide juicy morsels for a wide range of predators. Most species inhabit silken retreats hidden under stones, leaves or bark, and only rush on to their webs when an insect becomes entangled. Some, however, actually live on the web itself – an exposed and rather dangerous situation. It is not surprising, therefore, to find that numerous devices have been evolved to help them avoid predatory attack, some of which have already been mentioned in Chapter 3. A species of *Gasteracantha* from the Andaman Islands protects itself by collecting and binding together the insects it has sucked dry to form two heaps, coloured like itself. A bird or another enemy has thus only one chance in three of catching the spider in its first attack – and it will not get another chance. Everyone who has collected spiders knows that an orb-web spinner immediately drops to the ground when disturbed. As the web is usually woven in dense, often prickly, vegetation, the spider disappears into perfect safety.

Another spider *(Cyclosa)*, from India, makes a vertical string of pellets containing dead insects, seeds and pieces of leaf. It also spins four concentric ribbons around the hub of the web. These are white and extremely conspicuous, but the spider is lost to view in the complicated system of pellets, ribbons and anchoring threads. In Iraq, a species of *Uloborus* constructs a flocculent band across its web, drawn out here and there by the anchoring threads. The band is absent on the hub of the web, as this is where the spider rests with its first two pairs of legs thrust out in front and the last two pairs behind. The colour, size, position and immobility of the spider combine to make it appear like part of the band. A Burmese species of *Tetragnatha* places a withered stick in its web and presses itself against this. The spider is very long and thin, and its colour blends exactly with that of the stick.

Tooth & Claw

If the stick or web is touched the spider remains motionless, relying on its camouflage but, if it is further disturbed, it drops to the ground and is lost for ever.

The term 'stabilimentum' is applied to structures of ribbon-like or tufted silk, built into the snares of orb-web spiders. Most stabilimenta radiate outwards from the hub of the web and surround this region. To the human eye they are very conspicuous against a background of tropical vegetation or dark forest. In webs of spiders of the genus *Argiope*, for instance, this silvery white zigzag bands form a very confusing pattern that draws attention away from the spider. This, in any event, has been given an unusual and unspider-like shape because the zigzag bands appear to be continuous with its body, which has bright silvery markings.

It is not easy to decide whether such structures should be classified as deflective marks, camouflage, or disguise. Some may be one, some another, some may have a combination of functions, some may not have any defensive function at all. *Argiope argentata* builds a stabilimentum in the form of a white cross, and rests with two of its legs pointing down each of the diagonals. Experiments with wild birds as predators have indicated that they can use stabilimentum-like models to locate prey. It has therefore been argued that the stabilimentum is not an anti-predator device – although this does not necessarily follow. Instead it has been claimed to have a mechanical function. Since stabilimenta are often incomplete, it could be that each individual web differs in its mechanical properties, and that the spider needs some means of final adjustment which can be applied to the completed web after it has tested the effect of applying its own weight to the hub. Wherever the truth may lie between such contradictory views, an orb-web spider, resting on its snare, must be exposed both to predators and to solar radiation. Silvery, metallic coloration of the body, however, not only enables the spider to blend with the threads of the stabilimentum, but also reflects radiant heat and helps the animal to remain cool in strong sunlight. We may be sure that the relationships between orb-web spiders and their environments are extremely complex. That is why they have been discussed in so many different contexts.

Autotomy means self-amputation of part of the body: it is usually practised as a form of self-defence. For example, a lizard can break off its tail when it is seized by a predator, muscular action snapping a vertebra. Not only does the severed tail tear off in the enemy's mouth, but contraction of its muscles causes it to wriggle – thereby distracting the predator's attention while the lizard makes its escape. Many other kinds of animals are able to break off a part of the body when they are attacked. Sea-slugs and other marine molluscs autotomize their papillae which continue to writhe for some minutes after detachment and may, presumably, keep a predatory fish occupied while the mollusc crawls away: the papillae are regenerated later. They also contain defensive glands and, it will be remembered, nematocysts obtained from the sea-anemones on which these molluscs feed. Thus, if the warning colours of the papillae fail to deter an enemy, they may keep its attention diverted from the sea-slug long enough for this slow-moving animal to escape. Related molluscs autotomize and regenerate their tails, or even parts of the mantle, under similar circumstances.

Some anchoritic bivalves which live buried in sand or mud can regenerate a protecting siphon if this should be nibbled by a fish. Tube-dwelling fan-worms also have considerable powers of autotomy and regeneration. When a lug-worm *(Arenicola)* projects the posterior part of its body above the surface of the mud to defaecate, it immediately becomes vulnerable to predation. The hind region of the worm contains no vital organs, however, and can be autotomized and regenerated without harmful effects. Brittle-stars also readily autotomize their arms when attacked, and quickly regenerate new ones. Many arthropods, too, are able to autotomize their limbs and other appendages in defence. These are regenerated at the next and subsequent moults, but it may take some time before they reach full size. That is why crabs and lobsters, for instance, are not infrequently to be found with an unnaturally small leg or claw. Its predecessor will have been autotomized, and the replacement has not yet attained proper dimensions. Some social bees, wasps and ants autotomize their stings when they use them in defence of the colony, but in such cases, as already pointed out, there is no regeneration, and the insect dies for the benefit of her sisters which contain the same genes as herself.

Although larger tropical centipedes, or scolopendras, deliver painful venomous bites and secrete repugnatorial fluids, they are eaten with relish by many lizards, snakes, birds or mammals that are quick enough to catch them. I once kept a number of Ghanian specimens in a glass vivarium containing moist humus. One evening, I noticed that the humus had become rather dry, so I sprinkled some water on it. A few drops fell on the centipedes, which had already emerged from their hiding places and were engaged in their customary nocturnal activity. The falling water created a tremendous disturbance among them and, as they darted for cover, one of the animals autotomized a back leg, which lay apparently stridulating on the humus. The sound was audible from a distance of over a metre, and persisted for about 40 seconds until all movements of the leg had ceased. Subsequent experiments showed that sound production does not take place in attached legs, but only in the back pair after autotomy.

The creaking sounds which accompany the movements of the autotomized limb are doubtless an advantage to its former possessor, and distract the attention of an enemy even more than would the spontaneous movements of a silent leg, but this has not yet been proved. The long-legged scutigeromorph centipedes are also known to autotomize their legs when attacked. A vigorous contraction and relaxation of the muscles of the severed limb then takes place, accompanied by loud creaking sounds. The sound is produced by detached legs only. As they leap about, squeaking, they detract an enemy's attention from their owner, who slips away silently in the dark.

In no case has any stridulatory organ been recorded in the legs of a centipede, but the end legs of scolopendromorph centipedes, like all the limbs of *Scutigera*, are furnished with unusually long apodemes. (In certain parts of the arthropodan body internal projections, known as 'apodemes', serve for the attachment of muscles and the support of the brain and other organs.) I therefore suggest that the presence of these large apodemes, extending from the end of one segment far into the next, would impose a considerable strain on the cuticle near the point of attachment when the limb is flexed near its articulation. This strain, resulting in deformation of the cuticle in the region where it is weak, may well be accompanied by a creaking sound as the limb is bent and straightened – just as a plank of wood will creak if twisted about its axis.

In the same way that the attacks of birds against butterflies are often deflected from the vulnerable body to conspicuous eyespots on the wings, so the attentions of predators of lizards can be deflected to the tail. This is not only brightly coloured, but can be autotomized without much harm to its possessor. Quite apart from the fact that many lizards have brilliantly coloured tails which direct the attack of an enemy to a part of the body that can be dispensed with and replaced, autotomy of the most conspicuous part of the body distracts the attention of the predator so that its prey will have a good chance of escape.

An interesting parallel is seen in the sacrifice of its latest meal by a gull or tern when pursued by a skua. Skuas are pirates which relentlessly chase other sea birds until they are forced to vomit. Although not much larger than a herring-gull, the great skua has a bullying and aggressive temperament, and soon dominates its victim. At the same time, it is surprisingly quick and agile, swooping to catch the disgorged food before it reaches the sea. Since skuas can also kill and devour other sea birds, the loss of a meal seems a small price to pay for escape. It is easy to see how the habit of deflecting attack by, as it were, throwing the cargo overboard, may have developed as a ready means of escape.

Lizards fall into two groups with respect to the use to which their tails are put. In the first groups, self-induced autotomy is followed by regeneration while, in the second, the tail is normally retained and there is little or no regeneration following its loss. Within these general categories, lizards tails exhibit a wide diversity in form and function. Even among related species, the tail may be used for balance when climbing trees, for storing fat reserves, for sexual displays, as defensive weapons when used as a whip, or to deflect the attacks of predators. Where tail autotomy is most important for escape, regenerated tails are as large or even larger than the originals, but species that use their tails for other purposes do not regenerate them to any marked extent. When loss of the tail significantly reduces efficiency, as it would to a chameleon, selection favours retention because the cost of autotomy is too great. At the same time, because tail loss is an improbable or even fatal event for these reptiles, there has been no strong selection for tail regeneration. Thus, the functional nature of a lizard's tail, and the probability of its loss, are important in determining whether the tail is adapted for autotomy or for retention.

Small rodents, particularly dormice and fieldmice, are equally capable of shedding part of their tail when this is grasped by an enemy. Whereas the lizard tail drops away as a result of a clean break in the middle of one of the vertebrae, a mouse only sheds part of the tissue in which the tail is encased. The tail is not shortened but the enemy only gets its outer cover. Regeneration does not occur, however, and the denuded part of the mouse's tail withers away in a few days and drops off.

Of all the defensive activities found in the animal kingdom, two are seldom likely to occur simultaneously because they are apparently self-contradictory. These are retaliatory attack – with the aid of weapons such as teeth and claws, spines or stings – and withdrawal to a protected retreat. How can an animal launch an attack and, at the same time, escape down its hole? Among the few that have solved this problem, apart from spitting cobras which spray venom in the eyes of an intruder while they make their escape, are certain freshwater crabs *(Potamocarcinus)* which effectively combine both types of defence. An otter hunting along freshwater streams in Central America may make a swift lunge at a retreating crab, but will immediately retire – uttering high-pitched distress calls – with the claws of the crab embedded in its face or paws, or attached to the loose folds of skin beneath its neck, while the crab escapes into a hole or fissure. After counter-attacking with its claws, the crab autotomizes these weapons and escapes while the predator is still recovering from the shock!

A crab's claws have important functions in feeding, social signalling, and fighting so, although attack autotomy is a successful form of defence, it is usually only employed as a last resort. At the approach of a predator, such as an otter, raccoon or opossum, a freshwater crab will stridulate and raise the front of its body with the claws extended laterally. This display increases their apparent size, and reveals the striking coloration of the inner surfaces and of the bright red fingers. Only if the enemy, undeterred by such threatening behaviour, presses home his attack, will the claws of the crab be autotomized.

It is a popular misconception that octopods and cuttlefish use their ink to make smoke screens, as naval vessels did in World War I. In fact, the ink is discharged as a black blob. which does not diffuse into the surrounding water until long after the

Bluff, Death Feigning and Deflection of Attack

emergency is over. When an octopus or squid is attacked by a large fish, for example, the prey immediately retreats by jet propulsion. A blob of black ink then suddenly appears to attract the predator's attention. At the same moment, the prey instantaneously changes colour and darts off in a different direction. A combination of the sudden appearance of a black ink blob, combined with a change in colour and direction of movement provides a highly effective means of escaping any predator's unwelcome attention.

By a combination of bluff, death feigning, or deflection of attack to the wrong place, many animals avoid being killed and eaten by their predators. Even when a portion of the body is autotomized, it is not a vital part – except for the stings of some social insects, in which case other members of the same colony, with the same selfish genes, benefit. A predator may get some small reward for its efforts, like the tail of a lug-worm, or the arm of a brittle-star; but it is usually left only with a distasteful papilla, a poisonous sting or a horny claw – in each case the most noxious part of the prey. This may well suffice to teach it to be less rash in future, and not again to attack that particular kind of prey. Warnings and threats are useless unless the potential predator has a reasonably good memory. Aposematic coloration, mimicry, bluff and so on are no defence against insects, arachnids and more primitive enemies.

11 Withdrawal, Escape and Flight

Primary, or passive, defence, which operates at all times, whether an enemy is present or not, can never be perfect. It is almost inevitable that, sooner later, the prey will be discovered – either by chance encounter or as a result of systematic searching by the predator. When this happens, secondary defences are put into operation. The first responses of prey animals to a predator are sometimes merely an exaggeration of the primary defence. A newly hatched herring gull will lie motionless, pressed to the ground and relying on its wonderful camouflage. Stick-mimicking mantids stretch their legs so that the resemblance to a twig is enhanced, while aposematic animals display their warning colours and sounds even more dramatically than usual. Some examples, such as deimatic behaviour, death feigning and deflection of attack, were discussed in the previous chapter. Others are seen in withdrawal to a prepared retreat, flight or retaliation with weapons of various kinds including poisons, spines, horns, teeth and claws. Primary defences such as armour, spines and repugnatorial chemicals may also be employed actively.

Many anachoretic animals which emerge from their burrows or retreats in order to feed, mate or disperse themselves, respond to the presence of a predator by retiring rapidly down their holes. Fan-worms retreat into their tubes if a shadow crosses them, trapdoor-spiders back into their holes when the ground vibrates and shut the door, rabbits disappear down their burrows when disturbed, while hermit-crabs retire into the shells they have adopted as homes. The pearl-fish *(Carapus)* has a less salubrious refuge. When disturbed, it disappears, tail first, through the anus and thence into the body of a sea-cucumber. The pearl-fish is partially parasitic, feeding on the reproductive organs of the sea-cucumber which it first enters when quite young. Although it

Withdrawal, Escape and Flight

also feeds on shrimps and other small crustaceans, the pearl-fish regularly returns to its host for protection when disturbed by a predator.

Numerous other animals, which are not anchorites, also withdraw to prepared retreats when danger threatens. Mammals as different from one another as the platypus, the fox, the rabbit and the armadillo, the jerboa and the ground-squirrel, all construct subterranean burrows to which they retire for safety. In many cases, an emergency exit is constructed so that the mammal is not trapped when a predator is small enough to get in. The weasel that follows a rabbit down its burrow may well find that its prey has escaped through a second entrance!

Many animals are able to close the entrances to their holes after taking refuge inside. Hermit-crabs do this with an enlarged claw, while marine worms and snails often have a calcareous 'operculum' or shutter to keep enemies out. The African ground-squirrel *(Xerus)* blocks up the mouth of its burrow when it retires for the night, scraping the earth loose with the forepaws and shoving it out behind with the back feet. The loose soil is pushed towards the entrance, but not beyond, so that it accumulates to form a barrier. The fairy armadillo *(Chlamyphorus)*, although only 14 cm long, is not protected by scales. Unlike

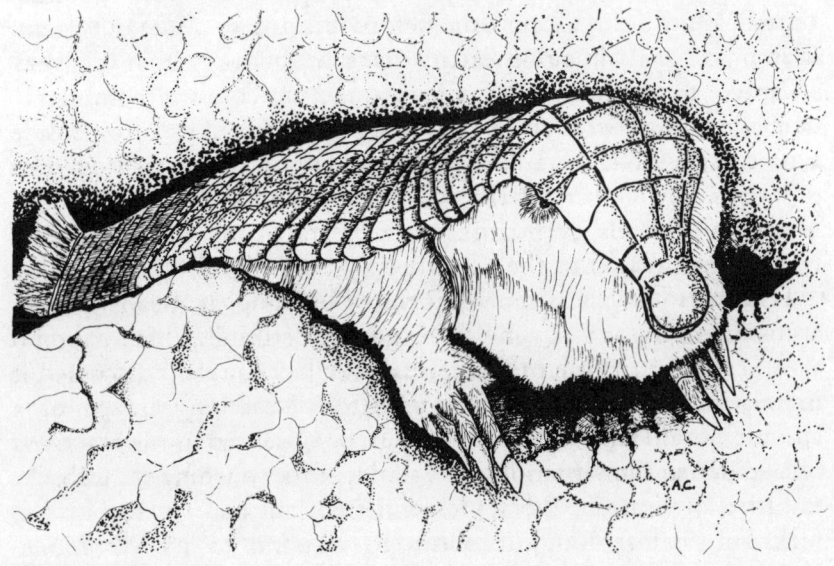

33. Fairy armadillo.

most of its relatives, it is furry and only partially covered with a few scales attached to the backbone. If its sole defence were to roll up, it could easily be killed by predators. When frightened, however, it quickly burrows into sandy soil, its rump protected by a bony shield and the armoured tail.

Various mammals which rest underground block the entrance when they go out to feed, leaving their babies inside safe from predatory enemies. Rabbits do not construct bolt holes in their nesting burrows, but always close the door when they leave. Sand-wasps *(Ammophila)* provision their subterranean egg cells with caterpillars, and then close the hole with pebbles and sand. They return from time to time to bring more food but, before finally leaving, the entrance is always blocked and camouflaged.

Too ready a withdrawal into a prepared retreat can, however, be disadvantageous, because the stimuli that induce an animal to do so may appear when there is no predator around. More intelligent creatures would probably recognize a false alarm very promptly, and not waste much time over it, but worms and other tube-dwelling marine invertebrates, for instance, are not very intelligent, and their ability to recognize predators is poorly developed. Since they have no means of detecting whether the enemy is still in the vicinity or not, they have to remain in their tubes or holes, unable to feed or to perform other essential activities, for quite a long time before it is safe to come out again. It is possible to imagine earthworms or snails, which are very sensitive to vibrations of the ground, becoming completely neurotic if they were to live on a railway embankment or beside a motorway. They would never have enough time to feed! But this does not happen. If the soil is caused to vibrate constantly, worms and snails eventually cease to show any response: they become habituated to the stimulus.

Habituation is the simplest form of learning. It means getting accustomed to, and ceasing to respond to stimuli which have no significance in the life of the animal. We have already discussed it in relation to disguise and a predator's searching image of a specific prey. Habituation can readily be observed in the reactions of a spider to the vibrations of a tuning-fork touching its web. At the first application, the spider will rush out and try to bite the fork but, before long, it ceases to respond to its vibrations. Habituation can be a source of danger to an animal, particularly

Withdrawal, Escape and Flight

in relation to traps set by Man: indeed, it is frequently exploited. After becoming habituated to the presence of a trap in its territory, an animal may lose its fear and go inside to eat the bait. On the other hand, if a tube-dwelling marine worm were to withdraw regularly simply because a frond of seaweed drifted back and forth above it, the unfortunate creature would probably starve. In fact, after repeated stimulation, the percentage of animals withdrawing in response to a particular stimulus begins to decline.

This might seem to imply that a persistant predator could always get its worm, but such is not the case. Animals that show habituation to simple stimuli – such as vibrations of the substrate or a shadow passing overhead – are remarkably slow to become habituated to a combination of stimuli. The ability of an animal to habituate is related to the stimuli provided by its predators. These are likely to cause several stimuli at the same time – shadows, vibrations, touch or chemicals – but a worm does not habituate to complex stimulation, except after very many trials. The speed with which habituation occurs is also related to the degree of danger. Decreased light intensity need not imply that a predator is close at hand, and habituation to this stimulus is rapid. On the other hand, contact implies that the enemy is about to eat the worm, and habituation to tactile stimulation is very slow indeed.

The usual response of an animal to a predatory enemy is to escape – according to its nature – as quickly as possible by running, jumping, flying, swimming, or merely by dropping to the ground. Many animals flee directly away from the enemy, like a sparrow chased by a cat, but the effectiveness of flight is often enhanced especially when there is no hiding place nearby, by jinking, zigzagging, looping or spinning. Most of the mammals and birds of open country follow a zigzag flight path when they are chased. This can be seen among creatures as different from one another as the field-mouse, vole, hare, ptarmigan, snipe, rhea and sandpiper; while the frequent unpredictable changes in direction made by gulls in the presence of a peregrine falcon have also been noted. Most people will have remarked upon the erratic leaping and jumping of sandhoppers disturbed by removal of the rotting seaweed under which they were hiding. Similar erratic movements are employed by fleas when hunted by their irritated hosts, and by water-fleas and other aquatic crustaceans when fish are after them.

Irregularity of escape behaviour is termed protean when it is sufficiently erratic to prevent a predator from predicting the future position and actions of its prey. It is a form of primary defence or insurance that occurs at all times and is not restricted to emergency situations. The flight paths of aphids, whiteflies, butterflies, flies, and other day-active insects usually follow contorted courses, whose protean elements are merely exaggerated during escape from predatory birds. Similarly, most small aquatic crustaceans swim in a characteristic jerky manner which becomes even more obvious when they are attacked by a fish.

Many water birds are especially vulnerable to hawks and eagles while they are losing height to land. Ducks and geese frequently descend in a swift, erratic, zigzagging and side-slipping manner. It seems more than probable that the crazy courtship flight of peewits *(Vanellus)* and godwits *(Limna)* may have survival value and compensate for the fact that courtship tends to make the birds conspicuous.

The essential feature of protean flight is a sequence of swift changes in direction. This is often reinforced by 'flash' coloration or colour change. The escaping cuttlefish provides a most striking example. At first it may be white with black spots, then it assumes a striped pattern – finally disappearing altogether, while the predator's astonished gaze is fixed upon a blob of black ink. Similarly, reef-dwelling fish perform rapid changes of colour when disturbed and, in some species *(Epinephalus)*, the display is rendered even more startling by a series of percussive sounds. In general, prey animals that employ the most protean changes in their flight behaviour are those most likely to escape. Moreover, sudden, quick movements are often quite frightening, and particularly so when they are unpredictably combined in a rapid series. Indeed, protean defence of this kind verges into bluff and startle behaviour.

Just as aircraft can be detected by radar, so ultrasonic echolocation is used by many animals to detect prey, find their way about or communicate with one another. Of all animals that use sonar, bats have been studied the most intensively. Not only can they fly at high speed in the dark, often through complex caverns or ruined buildings, but many species feed exclusively on small insects which are intercepted in mid-air. Lazzaro Spallanzani first experimented on the guidance mechanisms of insectivorous bats in 1794. He found that complete darkness or even blinding made no difference to a bat's ability to detect fine wires in its path. These, and similar experiments with blindfolded animals, have subsequently been repeated many times, and it is now known that bats make use of echoes from the ultrasonic sounds they themselves produce to locate and capture their prey. Cave swiftlets *(Collocalia)* and the oilbird *(Steatornis)* use sonar to avoid obstacles when flying in darkness, but do not hunt for moving prey as bats do.

So successful are bats in capturing flying insects, that they have become an important agent of natural selection. Consequently, it is not surprising to find that some of these insects should have evolved means of protecting themselves from attack. Many nocturnal moths possess 'ears', or tympanic organs, which are sensitive to ultra-sounds so that evasive action can be taken when the approach of a bat is detected. Noctuid or owlet-moths, in particular, may be captured on the wing, but are not taken easily. These heavily built insects, whose larval forms include cutworms, boll-worms, and other important agricultural pests, have a pair of ultrasonic tympanic organs on the third segment of the thorax, just in front of the waist, and behind and below the second pair of wings. When approached by hunting bats, they show a variety of flight patterns – loops, spirals and changes in speed and direction – in consequence of which they are able not infrequently to out-manoeuvre their pursuers. Analysis of 402 encounters between moths and feeding bats showed that, whereas approximately half of the moths that failed to react were caught and eaten, only 7 per cent of those that responded were captured.

In experiments using ultrasonic loudspeakers, it has been found that the insects which change their flight path erratically when they encounter sonar waves are mainly noctuid or geometrid moths and lace-wings, while many of the non-reacting insects are either hawk-moths (Sphingidae) or Saturniidae – the family which includes the emperor-moth and the giant silk-moths of India – which lack tympanic organs and are therefore probably deaf.

In addition to being able to detect ultra-sounds, some of the tiger- and ermine-moths, as well as species of a related tropical American family, also have the means of producing them. The exact function of these ultrasonic signals is still not clear. In most cases they are probably associated with unpleasant flavours and operate as warning signals to protect the moths from being eaten by their enemies: sometimes they are mimicking the sounds produced by distasteful insects. It seems improbable, however, because of their frequency, that these insect sounds can serve to jam the sonar of predatory bats although, as previously suggested, this is not impossible.

Animals do not detect approaching enemies only by sight and sound. Among marine invertebrates, smell is often the most important of the senses, both in hunting for prey and in detecting predators. Starfishes, which feed on mussels and other bivalves, are attracted towards certain chemicals produced by their prey. Strange to say, the common starfish of European waters is attracted chemically to mussels only between the months of November and May. During the remainder of the year it shows no such response, but whether this is because of a seasonal release of the attracting substance, or because the starfish has a lowered sensitivity between June and October, is not known. Scallops react in a striking way to the chemicals secreted by the tube-feet of various species of starfish. They either swim away vigorously, jump about in a protean manner, or rapidly close the valves of their shells. Larger starfish engender active locomotory escape responses in scallops more frequently than do smaller species which feed mainly on sea-urchins and mussels. Unable to harm adult scallops, which are quite safe once they have closed their shells, small starfishes seldom induce escape responses.

On land, the larvae of the pine looper-moth are often attacked by wasps – which show a significant preference for green, rather than for yellow larvae: the cryptic green caterpillars are probably more tasty than yellow ones. Furthermore, it has been found experimentally that wasps land more frequently on twigs smeared with the blood of green than with that of yellow caterpillars. This suggests that, unlike birds, wasps hunt by smell to a greater extent than by sight. Flesh-eating mammals stalk their prey upwind, otherwise they will be detected and cause it to bolt. This reaction is sometimes exploited by lions hunting in prides, and occasionally by other predators that operate in packs. As we

Withdrawal, Escape and Flight

saw earlier, it is not unknown for one predator to move upwind and drive the prey into the jaws of its mates lying hidden in ambush nearby.

Although flight is the habitual response of most animals to the approach of a predator, it does not usually take place until the enemy has reached a specified 'flight distance', which depends to some extent upon its size and speed, but varies according to circumstances. For instance, in Natal I once approached a blesbok to within 30 metres while on horseback. Had I been on foot, the antelope would never have let me get so close because its flight distance would then have been greater. Again, the flight distance of game animals to a vehicle is far less than to a man on foot. As an animal becomes tame, the flight distance within which it will not allow a person to approach is gradually reduced. Antelopes can tell the difference between a hungry and a fully-fed lion, and adjust their flight distances accordingly. If a predator approaches slowly, its prey will move off quietly, maintaining its flight distance. If the predator makes a dash, the prey responds immediately.

When an animal is approached to within a 'critical distance', which is much closer than the flight distance, automatically it must stand and fight. Hence the danger of following a wounded buffalo into thick bush: if the hunter gets within its critical distance, his quarry will certainly charge. Similarly, a cornered rat or weasel will fight ferociously. A caged animal regards the cage as its territory. That is why anyone foolish enough to climb into a lion's or tiger's cage is inevitably attacked.

As a predator approaches or oversteps the flight distance, the prey may react in various ways depending on its enemy's behaviour. If the latter starts a chase, the prey will move off at full speed. The outcome will depend almost exclusively upon the speed and endurance of both predator and quarry. Prey species are often faster than their enemies. African lions can reach maximum speeds between 50 and 60 k.p.h. but, with few exceptions, hoofed animals can run even faster. If, on the other hand, the predator does not attack, an ungulate, while keeping a more or less constant distance for its enemy, may give alarm calls and show a variety of additional responses such as 'stotting' or 'pronking', the stiff-legged jumping of gazelles, hartebeest and kob. It is especially characteristic of springbuck, from which activity these animals get their name. S. G. Cronwright-Schreiner described it as follows: 'The attitude of the pronking springbuck is very like that of a bucking horse. The head is lowered almost to the feet, the legs hang fully extended with the hoofs almost brushed together; this arches the back sharply and throws the haunches down, making the legs appear unduly long. In an instant the buck seems to spurn the earth as it shoots up into the air to an incredible height, perhaps straight up; for an instant it hangs arched, then down it drops . . . The buck seems scarcely to touch the earth when it bounds into the air again like a rocket, perhaps with a prodigious leap forward and as high as before; for a second you see it in the air, its mane up, its fan raised and opened in a sharp arch, the white patch blazing in the sun and the long hairs glittering, the legs and head all hanging in a bunch under the body; then it touches earth again, only to bound up once more at a sharp angle to one side, then straight up, then the other side, then forward, and so it goes on.'

The probable explanation of this behaviour is that the buck lives on the open veld where its enemies – lions, leopards, hyenas and lynxes – are unable to catch it by speed, but must rely on lying in hiding, rushing out and pouncing upon it. Sudden pronking is the springbuck's safeguard, and even kids, only a day or two old, will do it. In addition, such action affords warning to other members of the herd as it is conspicuous and can be seen from afar. Other conspicuous movements and postures, such as raising or curling the tail, and displaying the white underside are found in the more gregarious species and probably serve as signals to other members of the herd. Furthermore, they may also indicate to a predator that the prey is aware of its presence, alert, and therefore not worth chasing.

Antelopes graze on grass, or browse on leaves or shoots. Like most other even-toed ungulates, and in contrast to horses, tapirs and rhinoceroses which have an odd number of toes, antelopes are ruminants and chew the cud. Consequently, they are able to adjust their feeding times in such a way as to afford themselves maximum safety. When feeding, they masticate their food for a short while only, swallow it hastily, and then collect more. The food passes into the first compartment, or rumen, of their four-chambered stomach. While they are feeding, antelopes are continually alert, using their keen senses of hearing, sight and smell.

Withdrawal, Escape and Flight

When danger threatens, they hasten swiftly away: only when at leisure, and in a comparatively safe place, is the process of mastication completed. The food which has been swallowed is then regurgitated into the mouth and chewed thoroughly. When it has been swallowed for the second time, it passes through the first chamber of the stomach to the second, third, and finally, to the fourth chamber where most of the breakdown action of the digestive enzymes takes place.

Mammalian predators can be divided roughly into two groups – pouncers and coursers. Pouncers stalk or wait for their prey and attack it suddenly, while coursers catch their prey by chasing it until it is exhausted. Cheetahs are the fastest of land animals, and can reach a speed of up to 95 k.p.h., but if they cannot catch their prey within less than a minute, they become exhausted and break off the chase. African hunting dogs, on the other hand, may follow a gazelle for half-an-hour or longer, until it can run no more.

An ability to accelerate rapidly is of prime importance to both prey and predator, but it diminishes as body size increases. This is because the force exerted by muscle contraction is proportional to the square of the linear dimensions of an animal, its weight to the cube. This may limit the size of the prey that a predator can catch without using more energy than it would gain by eating it! In mammals, the weight of the prey is usually about one-third that of the predator though, of course, it can be larger. Initially, therefore, the prey may be expected to have an advantage because, unless taken by surprise, it can accelerate faster than its heavier enemy. Some predators compensate for this by accelerating only one part of the body – which can be projected extremely fast. Salamanders, frogs and chameleons catch prey with their long tongues. A chameleon with a body length of 12 cm, for instance, can project its tongue up to 20 cm, with an acceleration, about 50 g (fifty times the acceleration due to gravity), reaching nearly 20 k.p.h. The forelimbs of praying mantids attain speeds of up to 0.4 k.p.h., while the head of a striking rattlesnake moves at 18 k.p.h.

The larger animals of open country have, of necessity, evolved great speed and endurance. This is related to the absence of hiding places to afford concealment from predatory enemies, the need to travel considerable distances to obtain drinking water, and to migrate. Even birds, such as the ostrich, bustards and the secretary-bird *(Sagittarius)* are able to travel far and fast on foot. The significance of natural selection on the fauna of open country becomes even more striking when steppe and savanna animals are compared with those of thickly wooded regions. Many of the latter are adapted for life in trees, and have prehensile tails and feet, and well-developed claws or adhesive pads to assist them in climbing. Even ground-living species are usually smaller than their close relatives of open country when they have to force their way through dense vegetation. The African forest elephant is smaller than the savanna sub-species, forest buffaloes are smaller than the plains species, and the same relationship is found in leopards. The Royal antelope *(Neotragus)* of the coastal forest of West Africa is the smallest ruminant in the world, and the dwarf antelopes of the Cameroons and Semliki forest are not much larger.

The shape of a mammalian limb is related to its function, based upon the principle of the pendulum: the shorter the pendulum, the faster its rate of swing. In order to accelerate a pendulum clock, the weight or 'bob' must be raised. Now, if the weight represents the main locomotory muscles and their insertions, a long, fast-moving limb can be achieved by shortening the bones nearest to the body and lengthening the more distant ones. This is a functional explanation of the evolution of the limb of ungulates. In these, the humerus and femur are comparatively short, while the bones of the hand and foot are elongated. In horses, zebras and other equines, all but one of the toes has been lost, while antelopes and bovines have retained only two toes. In both groups, the centre of gravity is raised towards the body. The length of the equine stride is further increased through massive elongation of the third metatarsals, which form the 'cannon bones'. Concentration of the muscles at the nearer end of the limb not only has the same effect as raising the pendulum bob, but increases the angles of insertion of the muscles. This gives a more direct pull across the shaft of the humerus and femur. At the same time, the muscles are shorter and thicker, which increases their power and speed of contraction. In contrast, the limbs of moles and other burrowing animals are very short, particularly the more distal bones, which increases the mechanical advantage and leverage of the muscles, although the limbs only move slowly.

Withdrawal, Escape and Flight

There can be little doubt that the pressure of predation has been the main driving force in the evolution of aerial flight. Although only insects, birds and bats, like the extinct pterodactyls, are capable of sustained flight, several other animals can glide for considerable distances to escape from enemies. These include flying fishes, flying frogs, flying lizards, flying snakes and flying squirrels. Flying animals need to have well-developed sense organs – eyes in birds or a sonar system in bats – a delicate sense of balance and nervous control. Many arboreal animals, such as flying squirrels *(Sciuropterus)* and the flying phalanger *(Petaurus)* glide by means of a fold of skin, supported between the limbs on each side of the body, which acts as a plane. In the flying dragons *(Draco)* and related species of lizards that inhabit South East Asia, the body is flattened and extended sideways by a pair of large, wing-like membranes supported by five or six elongated ribs, which can be spread out or folded at will. In some species the flaps are brightly coloured, and are used in courtship displays as well as in flight. Although their soaring powers are not great, flying dragons look rather like butterflies as they dart among the luxuriant foliage with their bright wings opening and closing in the sunlight. The flying geckos of South East Asia have permanent fringes along the flanks and tail. In combination with webbed toes, these enable the animals to parachute gently from tree tops, like flying frogs do. At the same time, the fringes break up the outline of the lizards, making them inconspicuous against the bark of the trees upon which they rest.

The golden tree-snake *(Chrysopelea)* of Borneo is another reptile which, when disturbed leaps from the branches and descends obliquely through the air. To sustain itself during its glide, the body is held rigid and flattened, while the underside is curved inwards, so that the creature rides on a cushion of air. The snake uncurls itself just before landing, becoming supple and rounded so that it is ready to slip safely away into the vegetation without delay.

Reptiles are more numerous and successful in the tropics than they are in cooler regions; and flight is an adaptation to life among trees. Consequently, flying reptiles are only to be found in tropical rain-forests. The forests of Brazil and the Congo are smaller than those of Asia. They contain fewer flying reptiles, and none so adept as the flying dragons. Flying reptiles are found mostly in South East Asia because that is where the largest areas of tropical rain-forest occur.

Flying frogs *(Rhacophorus)* glide with enlarged webbed feet which sustain them in prolonged leaps from one tree to another. Flying fishes *(Exocoetus)* and the flying gurnard *(Dactylopterus)* use expanded fins as wings. *Exocoetus* reaches a length of 45 cm: as it leaves the water, its swimming speed is only about 5m/sec, but the pectoral fins are expanded and cause the front of the body to lift until only the ventral lobe of the tail is immersed in the sea. This then vibrates rapidly with a sculling action, so that, within a second or two, the speed of the fish is increased to over 20m/sec. The pelvic fins are then expanded too, and the flying fish becomes airborne: it may glide for 350 metres or more at a height of about 30 cm above the sea. Gliding is probably assisted by a cushioning effect of the air, as is experienced by aeroplanes flying at low altitudes.

Adaptations for flight have been evolved on at least thirteen separate occasions among mammals, probably initially as a means of escape; but only the bats have attained true powered flight. In these, the wing membranes are extended between the body, the limb bones and four elongated fingers, leaving the thumb free for crawling and climbing. The entire wing membrane of pterodactyls was supported by the little finger – which was enormously elongated – while the bird's wing consists of feathers attached to three digits which are more or less fused together.

Insects are the only animals whose wings do not represent modifications of pre-existing limbs. Instead, they are developed from lobes on the sides of the thoracic segments of the body. Some authorities suggest that winged insects may have evolved from small forms carried by wind into the upper air. Those with the largest lobes would have been best able to glide in a controlled fashion, and were therefore the most successful. It has also been suggested that flying insects may have evolved from roach-like, flightless ancestors. In prehistoric Carboniferous times, these were preyed on by the ancestral spiders: they escaped by jumping from the tree-ferns on which they lived, and glided to safety on their thoracic lobes. When their prey took to the air, spiders evolved webs to trap them in flight. Finally, it has also been postulated that wings may have evolved from the abdominal

Withdrawal, Escape and Flight

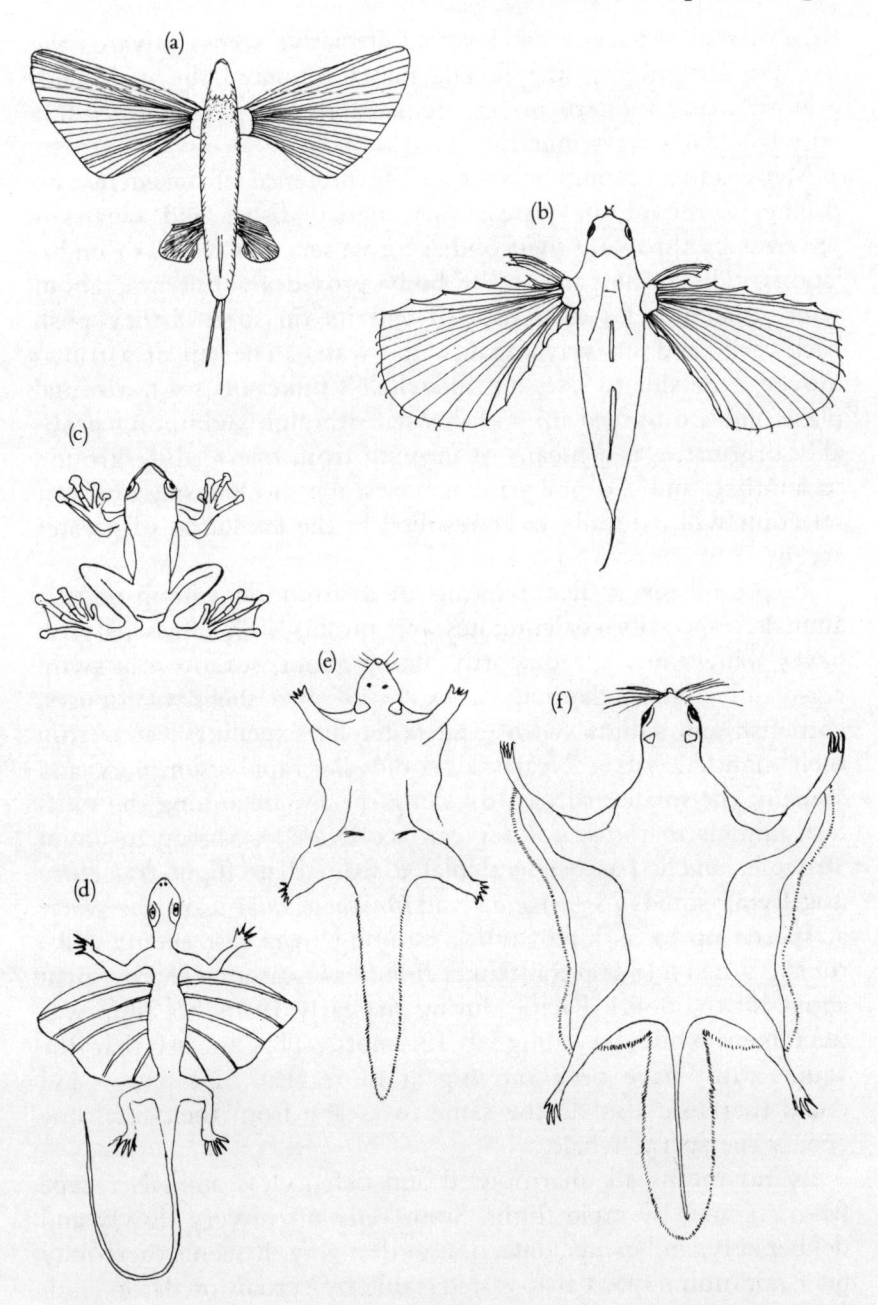

34. (a) Flying fish *(Exocoetus)*; (b) flying gurnard *(Dactylopterus)*; (c) flying frog *(Rhacophorus)*; (d) flying dragon *(Draco)*; (e) flying phalanger *(Petaurus)*; flying squirrel *(Sciuropterus)*. (After R. S. Lull. Not to scale.)

flaps of male insects which wafted attractive scents towards the females. Primitively, the female insect mounted the male and, even in many modern insects, females are often wingless while only the males are winged.

Many aquatic animals react to the presence of an enemy by fleeing. Some of the fastest swimmers – fishes and whales – progress by throwing their bodies into a series of waves or undulations. The front part of the body provides a fulcrum, about which the muscles turn the tail and its fin so that they push backwards and sideways against the water. The tail of a fish is moved from side to side, but the whale's flukes lie in a horizontal plane and are moved up and down. Although swimming probably originated as a means of moving from one feeding ground to another, and did not arise as an escape mechanism, predator selection will naturally have resulted in the evolution of greater speeds.

Jet propulsion is also a means of locomotion among aquatic animals, especially coelenterates and molluscs. Whereas jellyfish never achieve any speed worth talking about, scallops can swim reasonably fast by flapping the valves of their shells. Octopuses, cuttlefish and squids swim even faster by expelling water from their mantle cavities. Nearly all squids are rapid swimmers, and some of the smaller surface-dwelling species are among the swiftest animals in the sea. They can accelerate at 3 g up to about 10 k.p.h. and it has been calculated from their flight trajectory that flying squids *(Stenoteuthis* and *Onychoteuthis)* leave the water at speeds up to 55 k.p.h. when escaping from fast enemies like tunny. When a 15,000 ton tanker *Brunswick* was attacked by giant squids in the South Pacific during the early 1930s, the ship was overtaken while travelling at 12 knots. This means that the squids must have been moving at more than 32 k.p.h. – and could therefore also do the same to escape from their fearsome enemy the sperm-whale.

By no means all unarmoured and defenceless animals escape from enemies by rapid flight. Some only move very slowly and deliberately, and escape detection in that way. It seems there may be a minimum speed that is detectable by certain predators and, until that speed has been reached, they are not aware of the presence of their prey. Even we do not notice an animal moving very slowly as easily as we detect one moving quickly. The Brazilian two-toed sloth *(Choloepus)*, for instance, moves so sluggishly, hanging beneath the branches by means of its large curved claws, that it is almost unnoticed – especially as its fur is coloured green by minute algae growing in grooves along each hair. Similarly, the deliberate movements with which a chameleon approaches a fly are so slow that the prey fails to detect its predator. (At the same time, of course, enemies that might otherwise attack the chameleon cannot easily detect the animal in the trees where it changes colour to match its background.)

Even the simplest of flight reactions have frequently been altered and improved by natural selection. The animal whose behaviour has enabled it to escape the attacks of predatory enemies will be able to reproduce and pass to its offspring those instinctive traits to which it owes its life.

12 Horns, Teeth and Claws

Instead of fleeing from a predator that has overstepped the critical distance, a prey animal may proceed to threaten or even attack its enemy. This reaction is usually seen either in animals that are large and strong enough to put up a fight, whether individually or in a concerted group, or as the final defence of a wounded or cornered individual. Elephants, rhinoceroses and buffaloes can keep most predators at bay, including single lions, while some of the larger antelopes – such as oryx, roan, sable and greater kudu – may successfully repel lions, sometimes even killing one in a desperate fight.

Baby animals are often defended with astonishing ferocity by their mothers. No doubt this occurs when the female's critical distance has been infringed because she could not leave the baby after the predator has come within her flight distance. On one occasion a female zebra violently attacked a lioness which had killed her foal. The mare grabbed the lioness with her teeth at the nape of the neck and held on strongly. For a while, the lioness was unable to free herself. Then both she and the zebra let go simultaneously and separated, each with only minor injuries. Female wart-hogs successfully defend their young against cheetahs, while female gazelles commonly attack jackals and occasionally even hyenas, chasing their young. In such situations, they often strike out with the forelegs – weapons not used in intra-specific fighting.

In the evolution of the mammals, horns and similar organs have appeared many times, and have assumed a great variety of forms. For instance, the horns of antelopes are permanent structures borne on either side of the head. They consist of an unbranched bony core attached to the skull and surrounded by a horny sheath. These horns grow throughout life and are not

Horns, Teeth and Claws

shed: all male antelopes bear them, and some females too. The antlers of deer, on the other hand, are branched, bony structures which are shed every year. While the new replacements are developing, they are covered by skin and fur or 'velvet'. This is rubbed off when the antlers are full grown, by which time they consist only of dead bone. As a stag reaches maturity, its antlers become larger, and have more branches or 'tines'. Antlers are confined almost exclusively to males – except for the musk deer and the Chinese water-deer whose males are without them: in reindeer, the females have antlers as well.

35. Horns and teeth in central Africa.

The horns of rhinoceroses consist of tough epidermal fibres mounted on prominences of the skull. They have long been thought to consist of densely matted hair, but recent studies show that the filaments are not formed of hair. They are 0.3–0.5 mm in diameter, many-sided in cross section, laminated in structure, and hollow. They lack the 'cuticle' so characteristic of true hair and are held together by so little inter-filamentous horn that they fray easily and become worn. This is especially true of the Indian and Javanese rhinoceroses whose horns do not have sharp points like those of the African black and white species. Nevertheless, even they are formidable weapons which can do great damage to an adversary when a rhinoceros charges. Because they are in the habit of sharpening them on trees, the horns of rhinoceroses that live in forest regions of Africa assume long, thin, and dangerously projecting points, while the horns of rhinos that live in rocky districts become blunt and short from being sharpened on stones.

Horns are found in animals as distantly related to one another as the red deer and the *Mylogaulus*, an extinct rodent that lived during the Miocene period. They have evolved quite independently on more than one occasion even in quite closely related groups of animals, such as the ruminants and the rhinos, and assume a bewildering variety of shapes and sizes. In Africa, there are literally dozens of antelope species with differently shaped horns. In the oryx and duiker, for instance, the horns are almost straight; in the eland they are twisted; while the kudu, bushbuck, nyala, sitatunga and bongo bear spiral horns. The horns of the sable and roan antelope are long and curved; those of the buffalo and wildebeest are stout and curl forwards. Since many of these animals inhabit the same environment, it is difficult, at first sight, to see why they should have evolved so differently, or to recognize the advantages of such diversity.

The first stage in the evolution of horns and antlers may be represented by the development of small horns, used for intraspecific fighting in association with intimidating displays and defence mechanisms such as flight behaviour and a thick skin with heavy connective tissue lying beneath it. The skin of the okapi, for instance, is 5 to 6 mm thick in places and as tough as the rubber sole of a shoe, while the skin of the American goat-antelope was at one time used as breast armour by the coastal

Indians of Alaska. In the second phase of their evolution, horns served not only as organs for piercing and bruising the body of a rival in the establishment of a dominance hierarchy, but also functioned as guards to block an opponent's attack and make it ineffective. Finally, they have become adapted for binding the opponents together during pushing and wrestling contests, or may be used merely for display purposes.

Among the sheep, goats and oxen, certain groups, following a different road of evolution, have developed horns that serve as battering rams in contests for dominancy among rival males. Male bighorn sheep, musk-oxen and bison charge one another with terrific force. Male mouflon sometimes take a run of 20 metres or more; the noise of their impact can be heard from a long way off and sounds like lumberjacks at work. After the clash, both rams step quickly back and charge again. In many such animals, a pad of hair over the forehead acts to some extent as a shock-absorber. At the same time, the skull has become strengthened, and the horns really massive. Indeed, the sizes of the horns of the various individuals in a herd parallels their social position and dominance order. Graded horn sizes serve as indicators of rank, allowing wild sheep to live in open societies so that strange individuals can meet and fit into the social structure with a minimum of energy wasted in combat.

When threatened by a predatory enemy, the normal response of an antelope is to flee. Only when cornered, or exhausted by the chase so that its critical distance becomes infringed, must it stand and fight. In such cases, the horns are often used for defence: it is not unknown for an oryx to stab an attacking lion with them. Modern research suggests, however, that horns have evolved primarily in relation to intra-specific combat, as mentioned above. The long, stiletto-like horns of the oryx, for example, are not only formidable weapons of defence, but are also beautifully adapted for ritualized fighting between rival males.

The functions of antelope horns are as diverse as the fighting ceremonials of their possessors. The males of some species strike their rivals from above: others thrust upwards from below – such tactics are especially common among short-horned animals such as nilgai and okapi. Antelopes with long, sabre-like horns push with them locked in the horizontal, not the verticle plane, sometimes from a kneeling position. In the chamois, the horns are used exclusively for ripping the body of the rival or enemy.

Fighting plays an important role in the reproductive behaviour of animals, and special weapons have been elaborated by which it is effected. The ridges and twists of antelope horns, like the tines and projections of the horns of deer, serve to lock together the heads of the antagonists as they wrestle in battle. It would reduce reproductive potential if rival males were often to kill one another, or even to damage each other in territorial fights. A wounded, though victorious male would be no match for his unwounded adversaries, would lose his dominant status and cease to reproduce. Consequently, natural selection seems to have resulted in the evolution of special weapons for sexual fighting that do not inflict too severe bodily harm upon the adversaries. Used differently, however, they may be extremely effective against predatory enemies. The rapier-sharp horns of the oryx, for instance, can stab a lion, but are never used as daggers in intra-specific fights: they merely serve to hold the heads of the opposing animals together as they engage in pushing matches.

Not infrequently, entirely different organs are employed for defence against predatory enemies from those used for intra-specific fighting. For instance, the horns of the giraffe are reserved for social encounters, whereas predators are engaged with the far more dangerous hoofs. It is quite possible that large antlers may act primarily as organs of display, rather than for fighting. Their great size in the most vigorous and healthy males may have an intimidating effect on opponents. Overt fighting would thereby be reduced, since the male with smaller antlers would be less likely to fight than if he were not thus intimidated.

The fact that antlers are found in males only, and are functional for only a very short period of the year, is a powerful argument against the hypothesis that they are related to defence against predation. Male deer lose their antlers in winter just when they are most harassed by their enemies. It has been shown that wolves attack about 12 moose for every one they manage to kill and that, even without their antlers, the latter are capable opponents. Deer are well known for the ability to defend themselves with their front hooves rather than with their antlers, while giraffes use both front and hind legs against attack. Flight is usually a better protection from predators for sheep, goats and small antelopes than is fighting: there is no evidence that wild sheep or mountain goats can, with their horns, fend off attacks by wolves. Furthermore, such animals rarely stray from places where a quick escape to precipitous terrain is possible. Only in those ungulates that face their enemies, and do not flee, are the horns normally used in defence against predators: an example is afforded by musk-oxen, which form a phalanx against attack by wolves and thrust at their tormentors with sharp, up-curving horns.

Reindeer and caribou are unique among the deer family in that antlers are not restricted to stags. Antlered females are, in fact, the rule rather than the exception in both the Old World reindeer and its close relative, the caribou of North America. This anomaly has occasioned considerable scientific controversy. The suggestion has been made that antlers, which are shed each year, may serve to dissipate heat by means of the flow of blood through the covering of velvet while they are growing during the summer. No evidence, however, has been obtained for this from physiological studies of reindeer antlers. Other hypotheses as to the function of antlers include pre-reproductive threshing of vegetation, erotic stimulation by means of antler contact between two males, and masturbation which, in red deer at least, is accompanied by drawing the tips of the antlers to and fro through ground vegetation.

After loss of the antlers, either from breakage, casting, or amputation, red deer stags become less effective in competition with rivals, lose their social rank in the bachelor group, and fail to secure hinds. While stags are living together in bachelor herds, the antlers appear to be their principal status symbol. During the rutting season, however, additional display symbols and characteristics are employed, and it is by the use of these for intimidation that stags without antlers (hummels) are sometimes able to be successful in mating. There is only slight evidence to support the idea that antlers function to attract hinds. None of these observations explains the possession of antlers by female reindeer and caribou, but their use in nature suggests that antlers permit females to compete with males in winter when available space is progressively reduced by snowdrifts. Antlers enhance the social ranking of females, and enable them to get a fair share of restricted food supplies. Reindeer and caribou calves, which also produce antlers, are extremely precocious and independent of maternal care during the winter. It appears, therefore, that the

Tooth & Claw

function of antlers in females is not associated with protection of the young, nor with the latter's quest for food after snow has fallen.

It seems probable that the horns and antlers of ungulates may have been evolved to subserve at least three different functions. First, horns that are similar in shape and size in both sexes often serve as weapons for defence against predatory enemies, as well as in intra-specific contests. We find them, for example, in oryx and roan antelope, gnus, oxen, buffaloes and bison (reindeer and caribou are exceptional in this context). Secondly, horns which are useless against predators function only in intra-specific fighting. They are not found in females – with the exception of reindeer and caribou – and occur, for instance, in the impala, reedbuck and waterbuck. Thirdly, horns which are never used as weapons have a ceremonial function only, like those of the giraffe. Such horns are not relevant to the subject of this book.

To some extent, the evolution of tusks and teeth parallels that of horns. The elephant's tusks, however, serve many purposes – digging, moving branches, resting the trunk, offence and defence. Elephants played an important part in the wars of the ancient world, and many of the armies of Asia and Africa were equipped with elephant corps. They were a terrifying weapon, especially useful against men who faced them for the first time and against horses untrained to meet them. Not only did they often carry soldiers in turrets on their backs, but the animals in opposing forces actually engaged each other in battle. The way in which they fought has been graphically described by the Greek historian Polybius in his account of the Battle of Raphia, and sheds light on the use of tusks in nature. The battle took place during the Fourth Syrian War, between Ptolemy III of Egypt and Antiochus II of Syria on 22 June 217 BC. The Egyptian army possessed 73 African forest elephants, while that of the Syrians had 102 of the Indian species. According to Polybius: 'The way in which elephants fight is this. With their tusks firmly interlocked and entangled they push against each other with all their might, each trying to force the other to give ground until the one who proves the strongest pushes aside the other's trunk, and then, when he has once made him turn, he gores him with his tusks as a bull does with his horns.

'Only some few of Ptolemy's elephants came to close quarters

Horns, Teeth and Claws

36. A variety of horns. (a) Red deer; (b) bushbuck; (c) muntjac; (d) pronghorn; (e) Chinese water-deer; (f) roe deer. The inverse relationship of tusks and antlers is illustrated in the series (a), (c), (e). (After V. Geist.)

Tooth & Claw

with their opponents, and the men in the towers on the back of these beasts made a gallant fight of it, lunging with their pikes (sarissas) at close quarters and striking each other, while the elephants themselves fought still more brilliantly, using all their strength in the encounter and pushing against each other, forehead to forehead ... Now most of Ptolemy's elephants were afraid to join in battle, as is the habit of African elephants; for unable to stand the smell and trumpeting of the Indian elephants, and terrified, I suppose, also by their great size and strength, they immediately run away from them before they get near them. This is what happened on the present occasion.' Although Ptolemy was using the forest sub-species of African elephant, which is dwarfed by the bush elephant and is smaller even than the Indian species, he finally won the battle.

It is interesting to compare Polybius' vivid description of the fighting of war elephants with a passage written by the great hunter Aloysius Horn, over 2,000 years later. 'Two bull elephants nearly full grown were having a fight on a sandbank. The two fighters did not charge each other but with head to head pressed each other back. There were great gaps in the sand caused by the weight and pressure of the fighters as they moved slowly in a circle. Now the younger one was forced head to ground and seemed fagged out and was bleeding from tusk wounds, and the larger elephant taking advantage of his position now forced his head up and jabbed him fiercely several times with his tusks.' If elephants can fight one another so effectively, it is not surprising that they are almost immune to predatory attack.

Although the primary function of teeth is clearly that of killing and eating food, they are also used both in defence against predators and in intra-specific fighting. Small teeth, like small horns, probably evolved first, in association with a heavy, thick skin, so that opponents were not incapacitated during sexual contests. Thus, wild boars develop very heavy layers of connective tissue, up to 6 cm thick, on the lateral sides of their bodies. During fights, the blows of the opponent are caught on this shield. Furthermore, the shoulders and sides of male wild boars are frequently impregnated with tree gum which matts the long hairs and thick underwool, forming an additional protective layer. Hippopotamuses fight savagely with their large teeth, but their

Horns, Teeth and Claws

bodies are protected by extremely tough skin, supported by thick layers of fat.

Despite their numerous uses, the tusks of elephants parallel the second phase of horn evolution, and serve both as guards as well as for bruising and piercing the bodies of rivals. They prevent any major sideways movement of an opponent's head in pushing matches, and prevent the latter from using his own tusks effectively. Like horns, teeth may also be used in sexual display. Teeth are used in defence against predators by a wide range of animals. Even quite timid species bite if they are in danger – the ferocity and courage of the cornered rat is proverbial. It may be worth while, at this point, recalling that normally inoffensive creatures will fight in self-defence when their 'critical distance' has been infringed and escape is no longer possible. Finally, of course, the teeth of poisonous reptiles are used both for the capture of prey and also in defence against predatory enemies.

Among the fossil dinosaurs of the Mesozoic era, from about 180 to 60 million years before the present, are to be found numerous examples of bizarre, seemingly non-adaptive structures that, until recently, have defied satisfactory interpretation. These include crests, spines, frills and other features which cannot be explained in terms of feeding or locomotion. Some of them may have afforded protection from the attacks of predatory enemies, and this explanation of their function has often been invoked, but in most cases it is not satisfactory because the structures themselves were not strong enough to have acted in this way. They are now believed to have been associated with combat and ritual display directed towards rivals of the same sex, and with advertisement and courtship displays to attract potential mates.

As we have seen in the case of horns and teeth, such structures seem designed to minimize injury to equally armed combats, and frequently serve as holding devices employed in head-to-head pushing or wrestling matches that test the strength of the combatants. Closely related species often show variations in the size and shape of such structures, which are related to differences in the style of fighting. Their dimensions may also be related to body size, strength and experience, so that the appearance alone of these structures can serve as a basis for dominance relationships. The deployment of weapons in threat displays towards rivals and in advertisement towards potential mates assumes great importance in the lives of most vertebrates. Since larger animals usually dominate small members of the same species, such displays often involve stereotyped postures that result in the greatest possible exposure of body surface to the rival. Structures such as dewlaps, manes, crests and 'sails' serve to enhance the appearance of size. They intimidate rivals and bluff enemies.

Because their teeth are such lethal weapons, crocodiles seldom use them during intra-specific fights, but allow threat displays to settle questions of dominance. The carnosaurian theropods *Acrocanthosaurus*, *Altispinax* and *Spinosaurus* also possessed large teeth. At the same time, elongated vertical spines which supported a web of skin were probably used in lateral displays to increase apparent body size. The thin, nasal horns of *Ceratosaurus*, the crests of the hadrosaurs, and similar projections on the noses of other dinosaurs, may have helped to heighten the formidable appearance of the head in threat displays. A similar explanation has been proposed for the great variety of horns and frills of the ceratopsians. The elongated nasal tubes of the later hadrosaurs may also have served as vocal resonators, while the more primitive species possessed a small nasal horn, presumably used as a weapon for butting rival males in intra-specific combat.

The double row of large dermal plates along the backs of stegosaurs have recently been shown to function as forced convection fins for the dissipation of excess body heat. At the same time, they have been interpreted as functioning to increase the apparent size of the animals in threat displays. Such combat and visual display structures, inferred for dinosaurs, do not differ in any substantial way from those seen in lizards and other living reptiles. Dinosaurs probably protected their eggs from predators, just as modern crocodiles and alligators do, and no doubt also guarded the hatchlings until these were old enough to look after themselves.

Beaks, claws, pincers, and other organs normally used for feeding can, like teeth, also be pressed into service, when necessary, as defensive weapons. We have already seen how certain freshwater crabs autotomize their claws in aggressive defence. One of the guiding principles of nature, engendered and finely tempered by natural selection, is economy. This is why the same multipurpose structures are so often employed at different times and in differing ways in offence, defence and for sexual display.

13 Co-operation

Instead of fleeing from a predator, several individuals may sometimes co-operate to ward off its attack, acting together as a social group. One of the attributes in which a social group differs from mere aggregation is that its members exhibit some degree of cohesion and co-operation; and mutual defence is one of the more striking examples of this. Some of the larger and better-armed ungulates, such as buffalo and eland, as well as zebras and elephants, sometimes collectively attack lions or other predators, and can even drive them away, whereas a single animal might well have succumbed. Musk-oxen, beset by wolves, similarly form defensive rings with the larger and more aggressive members of the herd facing outward while weaker individuals, particularly calves, are protected in the interior of the ring.

The attempts of predators to separate individual animals from a group are commonly counteracted by members of the group bunching together. This enhances their ability to defend themselves, and helps to confuse the enemy. Individuals of many ungulate species band together upon sensing a predator or when attacked. So do mongooses, elephants, rooks, jackdaws, starlings, long-tailed tits, gulls and other birds, many species of fish, and tadpoles. Certain insects – such as aphids – band together when threatened by ichneumon parasites. When pursued, schools of mullet *(Mullus)* split into two halves which swim away in opposite directions, so adding to the confusion of the predator. The tight formation flights of blackheaded gulls and terns, in response to the appearance of a hawk, produce the same effect by making sudden, synchronized turns. Attacking behaviour, or 'mobbing', is sometimes mixed with flight as a response to predators. Many gregarious bird species combine to prevent possible attack from a predator by mobbing it first. In the forests of Central America mixed flocks, composed of more than one species of bird, show adaptations in both colour and behaviour that function in maintaining the cohesion of the group.

The protean features of the escape paths of individual animals – zigzagging and so on – have already been mentioned. The effect is greatly enhanced, however, if many animals behave simultaneously in the same erratic way. If a large number of water-fleas are placed in an aquarium of goldfishes, for instance, fewer fleas are devoured in a specified time than if there had been less to begin with. Similarly, the efficiency of lizards in catching locusts declines with increasing density of the prey, and at high density lizards become incapable of catching any locusts.

Each individual of a group usually employs slightly different escape tactics. In a pheasant brood, for example, some of the chicks flee further than others; some move in a straight line, others zigzag or double back, while odd individuals may occasionally freeze into immobility. When baby ducks scatter, some of them surface closer to the enemy than they had been before, others further away. Random scattering on the approach of an intruder is a common reaction among animals that aggregate in tightly packed clusters. These include mice and other rodents, baby birds, fishes, squids, insects and baby spiders. The individual variability hinders the predator, and prevents it from predicting what any individual prey animal is likely to do.

Small birds do not only co-operate in defence by forming bands to mob and harass hawks and cuckoos. They gain a measure of safety merely by migrating together in large flocks for, if one of their number is taken, a predator may be satiated for long enough to allow the remainder to escape. If they were to migrate individually, the predator might have time to regain its appetite before the next bird passed by. The regular mass flights of sandgrouse to water from their distant feeding grounds in arid country are spectacular. Constantly calling to one another, the birds congregate with impressive synchrony at the watering place after feeding beforehand in places scattered over an area of hundreds or thousands of square miles. They probably gain mutual protection by being present at the watering place together in large numbers. The suggestion has also been made that any sandgrouse which has experienced difficulty in obtaining food will be able to follow other birds that fly off in a recognizably determined manner to good feeding grounds.

Social birds, whose colouring is mainly cryptic, often display bright recognition marks while in flight. These must be for the convenience of their fellows – even at the cost of conspicuousness to the individual. On the ground, they are protectively coloured: in the air they are conspicuous anyway, and the social advantages of showing recognition marks in flight more than compensates for any comparatively slight disadvantage to the individual. The bright blue patches on a mallard's wing, which are displayed when the bird takes flight, stimulate and 'release' flight behaviour among other ducks in the flock.

Animals of many different kinds not only benefit from migrating together but also merely from aggregating – especially at the time of reproduction – because predatory enemies in the vicinity become satiated long before they have eaten enough of their prey, or its eggs, to make serious inroads on the population. This explains the advantage to palolo worms of mass swarming; to oysters and other molluscs, fishes and amphibians of synchronized spawning. Synchronization also increases the chances of eggs meeting sperms and becoming fertilized.

In marine animals, reproductive activity is often synchronized by the tide and phase of the moon. A dramatic example is provided by the palolo worm *(Leodice)* which lives among coral reefs in the Pacific Ocean. In the breeding season, the hind parts of the worms become packed with reproductive cells. At the moon's last quarter, in October and November, they break away and float upwards to the surface of the sea where the genital products are discharged and fertilization takes place. The head ends remain alive in the reef and regenerate new tails. Spawning takes place at low tide on several successive days: as dawn breaks, great funnels of worms burst to the surface of the sea and spread out until the whole area is a wriggling mass of them, brown and green in colour. The individual worms are about 25–40 cm in length: they provide an annual feast for fishes. All round, sharks and other large denizens of the ocean cruise quitly along, gulping them in, but without causing any appreciable difference in their numbers. As the sun makes itself felt, a change begins to occur in the length of the worms. They start to break up into shorter and shorter lengths until, some three hours after sunrise, the entire surface of the sea shows nothing more than patches of scum. So regular is the appearance of palolo worms that the people of Fiji and other Pacific islands, who esteem them a great delicacy, know just when to prepare their boats and nets for the annual harvest.

The lunar breeding of a luminous worm *(Odontosyllis)* near the West Indies has unexpected historical interest. The worm spawns at the surface of the Atlantic during the night at the third quarter of the moon, and the shining light from the female attracts the male. The luminescent glow at the surface of the sea lasts only five or ten minutes: females appear first at the water surface and, along with the eggs, emit a stream of brilliantly luminous secretion. Males then rush in with short, intermittent flashes. Now, on 11 October 1492, at 22.00 hours, a mysterious light was seen from the poop of Christopher Columbus's ship, the *Santa Maria*, just in the region where this phenomenon normally occurs. It was compared with the flame of a small candle alternately raised and lowered. On that night the moon was one day from her third quarter, and it has been suggested that this may be a point of evidence of first importance towards settling the problem of the exact landfall of Columbus in the West Indies.

Similar examples of regular and synchronized spawning by different species of marine worms for defence have been reported from various regions. The larvae of oysters and other molluscs also appear in vast numbers at certain times of year because spawning occurs in midsummer, about ten days after the full or new moon. The grunion *(Leuresthes)* spawns in California during high spring tide. The little fishes come ashore on the top of a wave, lie for a moment on the sand laying their eggs, and drop back into the sea with the successive wave. The eggs develop in the dry sand, hatching when they are again covered with water at the next spring tide.

Gulls, penguins and other sea birds that nest together in large colonies likewise obtain protection for their eggs by the very large numbers that are produced at the same time each year. Rats, weasels and other local egg-eating predators can glut themselves to excess without in any way endangering the reproductive potential of the birds. If the mating period were not synchronized and extended over a longer period, predators would be able to take a heavy toll.

The most vulnerable period in the life of any animal, and especially a mammal, is normally shortly after its birth. Although young ungulates are unusually precocious compared with the blind, naked offspring of mammals that are born in burrows or defended by well-armed parents, some species are very much more active than others. The great majority of babies remain hidden and largely immobile for several days, but a few accompany their mothers from the moment they are able to stand. Each of the two systems includes a number of adaptations that enhance its effectiveness. The young that hide away are usually very cryptic, while those that follow their mothers can usually flee more quickly. Moreover, their parents are often better able to defend them. Protection is also achieved by synchronous breeding. Over 80 per cent of all wildebeest calves in the Ngorongoro Crater are born within an annual peak of 2–3 weeks, and survival is much lower in small herds than in larger aggregations. Low calf density, combined with lack of synchrony between small herds in different localities, results in high mortality from predation by spotted hyenas. In large aggregations, on the other hand, there are enough older calves to provide cover for the newly born young, making it much more difficult for hyenas to single them out. At the same time, synchronous reproduction is less important in large, than it is in smaller, herds.

Comparatively few instances are known in which protective resemblance is enjoyed not by a single individual but by a group of animals of the same species. Young caterpillars may, however, crowd together and collectively resemble bird droppings while, in Guyana, there is a membracid bug *(Bolbonota)* of which both the female and her nest are very conspicuous and easily mistaken for one of the numerous growths of a small white fungus common in the neighbourhood. Several species of bugs of the genus *Ityraea* in Africa aggregate in such a way as to mimic the inflorescence of a lupin. One of these bugs has both green and yellow forms, and observers have claimed that the green forms tend to rest at the top of the stem, with the yellow individuals below them. In this way, resemblance to an inflorescence is increased because the individual flowers open progressively. Related bugs, with similar habits, have been described in India and Sri Lanka.

Many aposematic animals have gregarious or social habits, and thereby add to their conspicuousness. This is typical of wasps, black and yellow saw-fly larvae, various aposematic bugs (e.g.

Graphosoma) and blister-beetles. Whereas most hawk-moth larvae are cryptic and solitary, aposematic species tend to aggregate in groups consisting of numerous individuals. Many distasteful butterflies that fly alone during the day roost together at night in dense masses. Although their warning colours are less apparent after dark, the form and scent of the sleeping assemblages are sufficient to enable predators to recognize and avoid them. Some aposematic fish swim very close together, making large and very conspicuous black balls in the water.

Whenever vertebrate animals group together, social behaviour soon evolves. There are three reasons for this. First, it may enhance the advantages of group living. European starlings, flying in a flock, react to a peregrine falcon by drawing close together so as to form a very dense flock, and by performing swift turns with a marvellous degree of co-ordination. The peregrine is able to catch a flying bird by swooping down at a speed that may exceed 240 k.p.h., and striking the prey with its strong talons. This method of hunting makes it vulnerable to collisions, and peregrine falcons do not swoop on to dense flocks of birds. So, from the point of view of an individual starling, the more that a flock tightens and bunches together, the greater is the chance that predators will be thwarted. Again, predatory enemies are sometimes deterred by the massed alarm notes of cedar waxwings in a group; leopards avoid combined attack by the formidable, dominant males of troops of baboons. Other social animals that confront their enemies collectively often gain protection thereby. Carnivores themselves also benefit from co-operative hunting, for social behaviour helps both predators and prey.

The second reason why social behaviour appears when vertebrate animals form aggregations is that the development of hierarchies and stereotyped interactions reduces the likelihood of diseases and parasites being transmitted among crowded individuals. Mutual grooming, too, assists in this. Finally, and most important of all, social behaviour enhances competition – so that dominant individuals with their superior strength, weapons, speed, agility and intelligence, secure maximum access to members of the opposite sex for purposes of reproduction. When Man had developed weapons, culture, and population sizes to levels that removed the threat of predation by other species, he simultaneously created a new predator which ensured that only the fittest individuals should survive – rival coalitions or groups of his own species!

No single anti-predator response has attracted more theories and produced fewer facts than the apparently altruistic behaviour of social animals that warn their neighbours of the approach of an enemy but, in so doing, draw attention to themselves. The problem is that, if the animals that give warning are those most frequently killed and eaten, natural selection would clearly operate to remove the genes controlling this type of behaviour. Although their evolutionary origin and functions remain somewhat obscure, alarm calls could perhaps benefit the individual that gives them if they were to stimulate similar behaviour in other individuals, making them, as with protean flight responses, all more difficult to localize and capture. Of course, even if the individual that gave the alarm call were most frequently killed, selection might still favour this kind of behaviour if the neighbours who heard the call, and thereby escaped, were relatives and therefore carried many of the caller's own genes. If the benefit to the individual could be reciprocated at some future time, then natural selection might also favour the evolution of alarm calls, but this is difficult to explain.

Yet another possible explanation could be that the giving of alarm calls is genuinely altruistic behaviour and can be explained in terms of 'group selection' – an hypothesis that is not regarded with much favour by most biologists. V. C. Wynne-Edwards, however, postulates that natural selection operates between competing groups or populations of animals, as well as between individuals. Groups whose members behave altruistically, suppressing their individual interests for those of the group, are likely to have higher reproductive success overall, and consequently contribute more to the next generation than groups whose members behave selfishly.

Thus, if there were two groups of starlings, for example, in one of which every bird fended for himself or herself while, in the other, warning were given to the approach of a predator, it could logically be argued that the second group would be the more successful in the long run and would therefore produce more progeny. That group selection must take place to some extent is self-evident. The problem really arises in explaining how it can operate fast enough in comparison with individual or gene selection, to play any significant role in evolution. Even so, many biologists deny that group selection occurs at all. For instance, Richard Dawkins argues strongly that alarm calls can be explained entirely in terms of 'gene selection'. A gene for giving an alarm call would prosper in the gene pool because it had a good chance of occurring, not only in the animal that might pay for its altruism by being eaten, but also in some of the other individuals thereby saved – including its relatives. In other words, this would be an example not of group selection, but of 'kin selection', and more of the same 'selfish' genes would be saved than lost by the apparent altruism.

Defence of the young by their parents is found throughout all classes of vertebrates. Parent cichlid fishes distinguish between inoffensive intruders and potential predators of their eggs and fry. The former are ignored, but the latter are chased away. This is easy to explain in terms of kin selection. Recognition of enemies by solitary individuals and by groups is often an instinctive reaction, released by a specific sign-stimulus which may be visual, chemical, auditory or tactile. By testing the reactions of various birds to models representing other species – including birds of prey – Konrad Lorenz and Nikko Tinbergen have demonstrated that the flight reaction in released by the silhouette of a bird having a short neck, and is characteristic of falcons, hawks, buzzards and eagles. A model representing a long-tailed hawk or a long-necked goose, according to the direction in which it was moved, evoked the appropriate flight reaction in young turkeys only when moving in the appropriate direction. As we saw in an earlier chapter, scallops and queens react to chemical substances produced by starfishes; tube-worms and mosquito larvae respond to shadows cast upon them; while scorpions are alerted by vibrations of the ground caused by the tread of a heavy animal and detected by sense organs on the legs, and probably by the curious comb-like pectines on their ventral surfaces described earlier.

Pheromones are chemical messengers which influence the behaviour of individuals other than those producing them. Female moths produce chemical scents which attract males of the species from a distance up to 2 or 3 km. Sex attrahents, such as this, are a particular kind of pheromone. Social insects have developed a system of pheromones which stimulate other members of the colony to follow in their footsteps. Among termites, trail pheromones are produced by sternal glands in the abdomen while, in ants, the trail is signalled by marking the substrate with particular chemicals, or by discharging volatile substances into the air. Trail substances may be produced by the digestive tract or by glands associated with the venom apparatus.

Social insects also release alarm pheromones by which they communicate warning of danger to one another. These are not only defensive, but can have a partly offensive role as well. Such alarm substances may be produced by any one of a number of glands. When excreted by an excited insect, the pheromone provokes typical alarm and attack reactions in other individuals from the same colony. The alarm pheromone secreted by an angry bee when it stings a mammal, stimulates other members of the hive to sting the enemy on the same portion of its anatomy.

Alarm pheromones are often produced at the same time as poisons and repugnatorial fluids. The contents of the mandibular glands are discharged through the jaws on to the enemy, thus marking it as an aggressor so that it can be immediately recognized by other members of the colony. At the same time, alarm is given by the diffusion of pheromonal vapours into the air. The sting apparatus of wasps and bees contains several glands which produce pheromones, while the poison gland itself often secretes alarm chemicals which are discharged with the venom. It is through the discharge of these pheromones that simultaneous attacks by entire swarms of bees are co-ordinated. Substances which act as alarm pheromones occur in the sting apparatus and in the mandibular glands of worker honey-bees. If some object with the alarm substance on it is placed at the hive entrance, the bees nearby become agitated: they assume characteristic tense postures, run about excitedly in circles, or make short dashes towards the object which soon becomes crowded with bees. The number of bees at the entrance to the hive increases as additional bees are alerted and come out from the interior. They do not, however, usually sting a stationary object – movement is an essential stimulus to elicit a stinging attack. Of the two alarm pheromones, the one associated with the sting is from 20 to 70 times more potent than the mandibular gland secretion, and the suggestion has been made that the primary function of the latter may be to deter nurse bees from giving food to larvae that have just been fed, and to repel robber-bees from other hives.

Alarm pheromones are mostly terpenoids in the case of ants, but their effects tend to vary. For instance, a pheromone may act as an excitant at high concentrations, releasing the typical aggressive posture and circular running patterns characteristic of alarm behaviour. When exposed to worker ants for long periods of time, it induces digging while, in low concentrations, the same chemical may serve as an attrahent to other members of the colony. In one species of ant, citral engenders attack while, in another, it causes alarm reactions.

Members of the same colony of ants may be stimulated by defence pheromones, and combine to smear their adversaries with a malodorous secretion, from the anal gland, which seems to have an irritating or toxic effect. They also produce characteristic odours which vary with the species, colony or caste, and which are important in recognition and inter-communication. Colony defence of a North American ant *(Pheidole)* has been studied in detail. This species employs a complex strategy against fire-ants *(Solenopsis)*, which can be divided into three phases. At low stimulation, minor workers recruit nest-mates over considerable distances, after which soldiers (major workers) take over the main role of destroying intruders well away from the nest entrances. When fire-ants invade in larger numbers, fewer trails are laid, and the soldiers fight nearer to the nest, along a shorter perimeter and in dense formation. If the invasion becomes still more intense, the *Pheidole* ants abscond with their brood, scattering in all directions.

Recruitment is achieved by a trail pheromone emitted from the poison gland of the sting. Soldiers can distinguish trail-laying minor workers that have just contacted fire-ants – apparently through the scent they have acquired – and respond by following the trails more aggressively than they do when recruited to sugar water. Soldiers are better at fighting than minor workers, and remain longer on the battlefield. Until recently, defence in ants has been regarded as employing relatively complex individual behaviour and simple colony organization. It is now known, however, that defence of the colony is at least as complicated and precisely organized as the more advanced and better-known forms of social behaviour – such as caste determination and recruitment.

Co-operation

On arrival at the scene of battle, soldiers become highly excited, snapping at the fire-ants with their powerful mandibles, and chopping them to pieces. Recruited minor workers also join in the fighting, but are less persistent and leave more rapidly. As a result, the proportion of soldiers tends to increase until, eventually, they out-number minor workers – although they constitute only 8–20 per cent of the total population in the great majority of nests. In close defence, a mass of soldiers, assisted by comparatively few workers, forms a short, tight, defence formation around the nest entrance. If retreat becomes necessary, minor workers begin picking up and removing eggs, larvae and pupae. The queen then joins the exodus, breaking through the fighting masses of soldiers and fire-ants, to flee on her own. If, at this stage, evacuation of the nest is prevented, the defenders are wiped out and their brood eaten by the attacking fire-ants.

Relatively defenceless animals may sometimes gain protection by aggregating with more powerful species. Baboons are formidable animals when together in a troop: they have been known to chase cheetahs, and even leopards from their territory. This behaviour may sometimes be exploited by more vulnerable animals such as impala which, on spotting a cheetah, will run towards a group of baboons as if to gain a sanctuary. The two species are frequently seen together and, although adult male baboons sometimes prey on new-born impalas, the protection that impalas may receive against other predators could well outweigh occasional losses to baboons.

Many other instances are known in which relatively defenceless animals obtain security through associating with more powerful species. Within its territory, a carnivorous animal, such as an eagle or fox, will have a nest or home which is surrounded by a protected area where prey is never hunted. This 'cease-fire' area is a protective device which automatically prevents parent animals from inadvertently attacking their own offspring. At the same time, however, it enables pigeons to nest with impunity beside a hawk's eyrie, and roe deer to be safe by the wolf's den.

In tropical countries, certain birds habitually build their nests close to the nests of venomous and aggressive insects such as wasps, bees and ants. For example, caciques *(Cacicus)* in South America regularly build so close to the nests of wasps that the homes of the insects and birds rattle against each other when the wind blows! That these remarkable nesting arrangements are concerned with the need for protection is shown by the fact that the species of stinging insects chosen as neighbours are always among the most virulent and vicious. Some caciques, as well as swifts and weaver-birds, gain protection from nesting near human habitations.

A curious association between predator and prey is developing among dogs and langurs, or leaf-monkeys, in Rajasthan. Langurs are common in Jodhpur and are often chased by dogs: they almost always escape by climbing trees or buildings. On the rare occasion when a dog catches a langur, other langurs attempt to help the victim by diversionary threats – gnashing teeth, waving their arms, or even by rushing in and clawing with their sharp nails. Dominant males are especially active, and occasionally achieve the release of the victim, but are ineffectual against large or particularly aggressive dogs. In certain localities, where langurs are sacred, they are habitually fed by the villagers and protected from dogs that attempt to molest them. In these places, dogs learn to leave langurs alone, not only because aggression is promptly punished with sticks and stones, but because docility is rewarded by food. The relationship occasionally goes beyond mere tolerance, however, and old, harmless dogs may even allow themselves to be groomed by langurs! In the particular conditions created by the local villagers, peaceful co-existence may also develop between dogs and protected antelopes – the chinkara and blackbuck. Vervet monkeys, likewise, sometimes groom duikers in Natal.

Many species of aphids and other plant-sucking bugs obtain protection from their natural enemies through an association with ants, from which both benefit. The ants are provided with food while the aphids are afforded shelter as well as protection from fungi, parasites, and predators such as ladybird beetles and their larvae, hoverfly larvae and so on. In order to obtain sufficient protein to provide adequate nutrition, aphids and other plant-sucking bugs have to imbibe far more sap than they would otherwise need, and the surplus is excreted as 'honeydew', a sugary liquid much appreciated by ants. It is not surprising, therefore, that a mutualistic association should have developed between these very different types of insect. In the absence of attending ants, most aphid species periodically produce droplets

Co-operation

of honeydew which are ejected either by kicking them off with a hind leg or expelling them by contracting the abdomen. When ants are present, however, they palpate the abdomens of the aphids with their antennae. This stimulates the aphids to exude droplets of honeydew and simultaneously inhibits them from ejecting them. Some myrmecophilous, subterranean aphids have even evolved a group of bristles in the anal region which holds the drop of honeydew while the ants imbibe it.

Aphid species possess various structural adaptations for defence against predators. These include repugnatorial 'cornicles' which emit wax which hardens rapidly; a dense coating of flocculent wax filaments which are secreted by groups of special glands on the body surface; a tough, sclerotized cuticle; and modifications of the legs for jumping. All these adaptations tend to be poorly developed in, or absent from, those species that are frequently associated with ants. This indicates that such associations have persisted for a very long time.

Many marine crustaceans obtain protection from predators by forming close associations with sea-anemones or sponges, and trading upon the unpalatable reputations of their partners. Certain hermit-crabs are almost invariably to be found with a sea-anemone adhering to the mollusc shell in which they have taken up residence. When obliged by growth to shift into more spacious quarters, the crab transfers its partner from the smaller shell, and holds it in position against its new residence until the anemone has become firmly attached. The sponge crab *(Dromia)*,

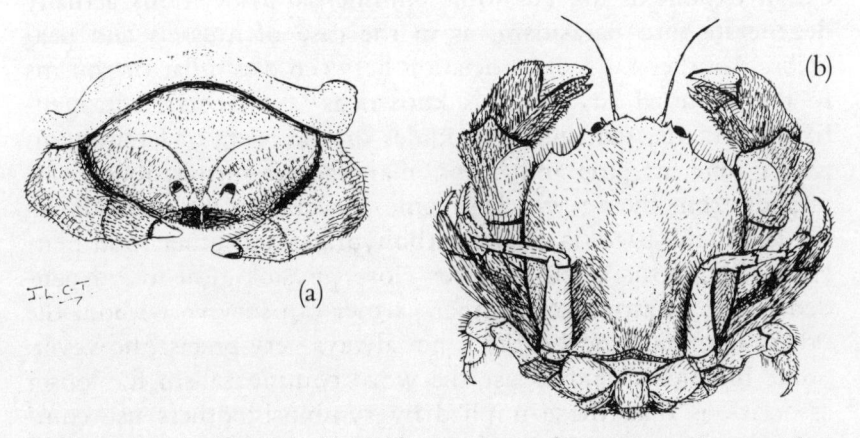

37. Sponge-crab *(Dromia)*. (a) Holding a mass of sponge on the top of its carapace; (b) seen from above.

213

already mentioned, has the last pair of legs shortened so that they can be used to hold a mass of living sponge in place over the back; while the coral-hunting crab *(Melia)* from Mauritius invariably grasps two anemones – one in each claw – employing them both for defence and for feeding. If the crab is attacked, it thrusts a polyp towards the enemy and wards it off with the anemone's stinging tentacles. On the other hand, if an anemone captures food, the crab takes the morsel with one of its legs and eats it. Damsel-fishes (Pomacentridae) of the Indo-Australian archipelago live in close association with large sea-anemones of the genus *Stoichachis*, among whose stinging tentacles they find shelter. Damsel-fishes are unable to exist without their anemones, upon which they depend absolutely for shelter from predators.

Mutually beneficial commensal associations are often found between tropical goby-fishes and the shrimps which inhabit their burrows. Actually, the shrimps seem to undertake more than their share of the work, removing sand and débris from the burrow of the fish and building, in front of the entrance hole, a large mound on which the goby sits. Apparently, they benefit by eating scraps from the goby's meals, but both partners take refuge down the burrow whenever danger threatens.

When two animals of different species live in association with one another, but without physiological interaction, the relationship is said to be 'commensal', Some examples were cited in Chapter 1. In commensal associations usually only one partner derives much benefit and, not infrequently, the other is to some extent exploited. Indeed, some commensal associations actually degenerate into parasitism, as in the case of mussels and pea-crabs. Another kind of association between dissimilar organisms to their mutual advantage is known as 'symbiosis'. Nitrogen-fixing bacteria, that inhabit nodules on the roots of leguminous plants, such as peas and beans, manufacture nitrogenous compounds from the air which become available to the plant. From the latter, the bacteria obtain carbohydrates and other food materials. Symbiosis often involves close physiological interdependence and, in rare cases, neither partner can survive without the other. Terminological usage is not always very precise, however. Some biologists tend to use the word commensalism for looser associations than those implied by symbiosis; others use commensalism for one-sided relationships, and symbiosis for situations which benefit both partners.

Certain fungi, which are parasitic on oak trees, live in close biological relationship with the scale-insects that feed on the sap of the oak leaves, growing over and adhering closely to them. Although some of the scale-insects are also parasitized, fungal growth is slow within their bodies so that they are able to survive throughout the dormant season – even though they are dwarfed and incapable of reproduction. The fungi are dependent upon the insects for some of their food and also for dispersal. At the same time, they afford protection to the scale-insects against their most formidable enemies – parasitic wasps – whose ovipositors are unable to reach the bodies of their insect hosts when the fungi are growing thickly. Thus, although some individual scale-insects, are sterilized or killed, the colony benefits from the association, as does the fungus. At the same time, both insects and fungi exist at the expense of the oak trees on whose leaves they live and feed. Close association such as this in which, on balance, both partners benefit, is clearly a symbiosis.

Another example of symbiosis is afforded by the plant-feeding insects and mites whose activities engender the formation of galls by the plants on which they feed. Many diverse organisms obtain shelter and protection from plants by inducing them to produce galls. Nematode worms sometimes cause huge swellings on the roots of plants, and many arthropods stimulate abnormal plant growth – excessive swelling of epidermal cells, and irregularities in the leaf surface producing sack-like outgrowths. In these, mites or aphids obtain relief from parasites and predators, and are protected from desiccation. The most remarkable galls are produced by psyllids, gall-midges, gall-wasps and saw-flies. These galls are abundant, colourful and often grotesque, while the organisms that cause them are usually small and difficult to identify.

The formation of galls was possibly one of the first of the activities of insects to be recorded by man although, for a long time, it was not realized that most galls are caused by living organisms. Theophrastus (372–286 BC) described their medicinal properties and Pliny, about AD 60, wrote of their rapid growth. But it was not until 1686 that their formation was first explained by the Italian physician, Marchello Malpighi in his famous work, *De Gallis*.

Tooth & Claw

Gall-insects attack nearly all parts of the plant – buds, leaves, petioles, stems, bark, roots and flowers. Very often they produce the same type of gall on different species of host plant. The inner walls of the gall are often richer in protein than the portion of the plant upon which it grows, and the mite or insect larva resting there finds an abundance of concentrated food. In general there are two types of galls: open and closed. Open galls are produced by aphids, coccids, psyllids and mites with sucking mouthparts which feed from the outside at first, but later cause leaves to fold and grow inward to form pockets in which they live in safety. Typical examples are the pear-leaf blister-mite, the witch-hazel cone gall-maker, and the elm cocks-comb gall-maker. Closed galls are made by the larvae of mandibulate insects, including beetles, moths, flies and wasps. Most of these, such as oak-apples, contain only a single larva, but a few, such as the oak hedgehog-gall, may house several larvae in separate cells or chambers.

Almost as remarkable as the gall structures produced is the fact that gall-insects restrict their activities to a comparatively small number of plant families and genera. More than 90 per cent of all the hundreds of known gall-wasps occur on two completely unrelated kinds of plants – oaks and roses – although the galls produced may be of very different types. Galls can be further classified according to their form and texture. Shallow-blister, or spot-galls are usually the work of gall-midges, as are the thicker spangle-galls that look like tiny saucers on the lower surfaces of leaves – especially on oaks and hickories. Button-galls, produced by gall-wasps, are thicker, as are oak-apples and the irregular woody galls found on oaks, rose and blackberry bushes. The galls found on flower heads are usually caused by gall-midges, but other kinds of insects and mites may also be involved.

The gall-making habit has undoubtedly developed independently in widely separated groups of animals, and may have arisen through the leaf-rolling responses of plants. The gall-insect lays its egg upon the host, or inserts it within the tissues of the plant. Upon hatching, the young larva begins to feed, and this stimulates production of the gall. There is a remarkable degree of specialization of both insect and host plant due to highly specific interactions from which both partners appear to benefit to some extent. For example, several species of ants live in hollow thorns of acacia trees. The thorns swell into gall-like growths – very conspicuous on the trees surrounding Nairobi airport, for instance – but this does not seem to be the result of anything the ants do to the tree. Some of the ants which live in such places are simply those which would move into any suitable cavity, however it had been formed, but others are found only in particular plants and appear to be completely dependent upon them. In addition to the acacias of thorn scrub in arid lands, the understorey trees of tropical forests often harbour ant colonies. It is not always clear what the plants get from the association – apart from protection from leaf-eating animals – but ant-plants are generally healthy.

At first sight it would appear that the relationship between poisonous plants and the insect specialists which can browse on toxic vegetation is a completely one-sided affair – no benefits whatsoever accruing to the plant. But a closer look at the situation reveals a few hidden advantages. Many plants have developed specializations for attracting insect pollinators. One of these consists of simulated insects feeding on the flower heads. For instance, a solitary purple flower in the centre of a white inflorescence looks, from a distance, deceptively like a fly; while flowers of the genus *Ophrys* have evolved structures that look like bees sucking nectar from them. The food plants of noxious insects with warning colours need no such device. Thanks to the sluggish habits of aposematic insects, which sit blatantly and for hours at a time on flower heads, no other source of attraction is necessary.

Another benefit which poisonous plants may acquire from acting as hosts to distasteful insects is the reinforcement of their protective odours, while the insects themselves benefit from acquiring the smell of the plants on which they feed. Thus, the powerful smells of certain plants act as warning deterrents to large herbivores, while the grasshoppers, butterflies and other insects which feed on these plants share their unattractive scents. The relationship is carried a stage further when ladybird beetles obtain repugnatorial chemicals indirectly from plants by preying on the aphids which suck their sap.

Defence against parasite attack has so far not entered much into our discussion, but a recently unravelled relationship between oropendolas, cowbirds and botfly larvae in Central and

South America is so fascinating that I cannot resist mentioning it here. Oropendolas *(Psarocolius)* are black birds with yellow tails, related to the orioles and blackbirds. They build hanging, flask-shaped nests in palms and other tall, spreading trees. Cowbirds *(Molothrus)* are brood parasites, like cuckoos, which lay their eggs in the nests of oropendolas, and are common around oropendola colonies which often, but not always, occupy trees inhabited by wasps and bees. Some oropendolas – usually those that inhabit trees which have wasps or bees living in them – discriminate against the eggs of cowbirds and eject them from their own nests. On the other hand, oropendolas that nest in trees away from stinging insects, do not react to the presence of a cowbird egg.

The explanation of this appears to lie in the fact that oropendola chicks tend to be infested by the larvae of botflies *(Philornis)*, which enter the birds' long, flask-like nests and lay their eggs directly on the nestlings. When these eggs hatch, the maggots which emerge from them feed on the tissues of the chicks, eventually causing their death. Oropendolas are protected from botflies by wasps and bees which kill or drive them away. Their babies are only vulnerable when the nests are situated in trees without bees or wasps in them.

Oropendola chicks that have a baby cowbird for a nest mate are nine times less likely to be infested with botfly maggots than if the 'cuckoo' were not present. This is because cowbird eggs hatch nearly a week sooner than the eggs of oropendolas, and the young cowbirds are extremely precocious compared with the chicks of their hosts. Their eyes open with 48 hours, while those of baby oropendolas do not open for 6 to 9 days. These adaptations presumably make the cowbird chicks more competitive than their nestmates. At the same time, however, young cowbirds snap aggressively at intruding objects, including botfly adults, eggs and larvae. Not only do they defend themselves against botflies but, incidentally, they defend their unrelated nestmates as well! Oropendolas therefore benefit from the presence of brood parasites if there are no bees or wasps nearby.

This is not, however, the end of the story, which is clearly far more complex and is not yet fully understood. Mites, too, play an important part in the ecology of the oropendola nest, and cowbirds, as well as oropendolas, behave differently according to whether wasps and bees, botflies and mites are present in the colony. In tropical forests, where the physical conditions of life are comparatively easy, inter-relationships between competing and symbiotic species become extremely complicated.

14 Predator-Prey Interactions

Some animals are particularly unlucky in having enemies of many different kinds. Grasshoppers, for instance, are eaten by a variety of spiders, lizards, birds and shrews – to mention only a few of their numerous predators. Other animals, more fortunate, are almost immune to most predatory enemies. Prey species that are subject to severe losses from attack by a small number of specialist predators may themselves evolve defences directed specifically against this limited array of enemies. For instance, the venomous scorpion-fish flees from a hunting octopus, but merely raises its spines when approached by large carnivorous fishes. The African ground-squirrel jumps vertically into the air if it hears a rustling noise in the grass. This is about the only defence that would be of use to it against a snake which was about to strike!

The meerkat *(Suricata)* of South Africa is a fawn-coloured, sharp-nosed mongoose, not much larger than a big rat. It is a gregarious species: small communities live in open country, often with ground-squirrels whose burrows they may take over – although they are quite capable of digging their own! Because of the open nature of the country in which they dwell, and because they are strictly day-active, meerkats must constantly guard against aerial attacks by hawks and eagles. They spend the day squatting on their haunches and scanning the skies for these aerial enemies. A hawk, even high in the air, will not pass unnoticed. The alarm is given by one of the meerkats, and the entire colony watches the predator intently until it flies away. If its shows any sign of contemplating attack, all the meerkats immediately take refuge in their burrows. This response to aerial predators is probably innate, and so is the tendency to look upwards. When a tame animal sees an aeroplane for the first time it is thrown into a panic and flees to shelter. After a few similar experiences, however, habituation occurs and flight gives way to vigilant watching, the alarm cry being constantly given for as long as the aeroplane remains in sight. Meerkats reared in Africa as pets often see hawks without ever being attacked. Nevertheless, they give alarm calls and remain alert and watchful in the presence of a hawk even when harmless birds have long ceased to evoke any response from them. This suggests either than an innate mechanism must prevent the waning of the response to the shape of hawk, even in the absence of any experience of attack, or else that meerkats only accept as harmless those birds that both fly low and yet fail to attack.

In human warfare, any particularly effective form of defence eventually elicits some new weapon of attack. Entrenched machine guns and barbed wire controlled the battlefields of Flanders until the advent of the tank restored mobility to the front. In its turn, this new weapon evoked the development of high-velocity anti-tank guns and, later, of bazookas, anti-tank rockets fired from fighter aircraft, and a host of other defences including

38. Meerkats *(Suricata)*. (Drawn from a photograph.)

guided missiles. Just as attack and defence interact in international holocausts so, too, do predation and anti-predator defences influence one another in animal evolution. First the prey begins to specialize in its response to the more dangerous of its predators. The predator then either develops some novel mode of attack to overcome the new defences of its prey, or else it finds something else to eat. I have already suggested that by pronking a springbuck may indicate to a predator that it is alert and therefore not worth the effort of chasing. White-tailed deer and other mammals have rump or tail patches which are used to signal to a predator and elicit pursuit. If the potential prey recognizes a hunting predator while the flight distance is sufficient to allow safe escape, display of the rump patch to the enemy may shorten the period of interaction. The predator may go off to look for less alert prey, or it may chase the displaying individual and fail to catch it. In either case, both the time and energy expended by the prey will have been reduced, and economy of effort achieved as a result of its interaction with the predator.

Evolutionary changes and adaptations evolve continuously over countless generations and many millennia – offence and defence continuously influencing one another – but the ebb and flow of confrontation are always present in nature. When different types of enemies are involved, an animal may develop different kinds of defence from each. It may also exhibit a sequence of both primary and secondary defences. For example, although mantids are voracious predators of other insects, they are themselves preyed on by lizards and birds. Not only are they cryptic, but they may show protective resemblance to lichen, bark, leaves or sticks. In some species, special attitudes of the legs enhance the resemblance. Some African species occur in two forms – green and brown. The proportions of these two colour variations alter seasonally, according to the predominant colour of the vegetation. Again, the early instars of several mantid species are ant mimics. Their secondary defence mechanisms include running, flying, death feigning, the secretion of repugnatorial fluid from the mouth, startle displays and flash coloration. Female mantids often defend their nymphs from intruding ants which are attacked agressively, but defensive attack against vertebrate predators is not common. Some of the numerous defences may be more effective than others against different kinds of enemy, but the number and variety of responses produced suggest that mantids must have to defend themselves against many different predators.

A similar diversity of defensive reactions is exhibited by many other kinds of animals. Dragon-lizards *(Goniocephalus)* of the Pacific region exhibit elaborate combinations of defensive responses. Their first line of defence depends on camouflage. Under attack, however, several secondary defences are invoked. These incorporate mouth gaping, inflation of the body, extension and contraction of the dewlap, the adoption of an extended four-legged posture, erection of the crest, tail lashing, death feigning and, occasionally, autotomy of the tail. Some species flee when escape is possible, dodging behind trees to avoid detection. When handled, they change colour to a bluish-green which contrasts strikingly with the brilliant orange of the gaping mouth. One species depends on bipedal flight, but may turn on an attacker and bite fiercely, emitting a guttural hissing sound as it does so.

As far as ant mimicry is concerned, Eric Wasmann (to whom reference was made in Chapter 9) provided an extensive survey of types of relationships that are found between ants and termites, and the spiders and insects that mimic them. These range from genuine guest relationships – where optical or tactile mimicry, or both, may play a part – to aggressive mimicry in which the mimic preys on its hosts. Examples include various spiders and heteropteran bugs, as well as staphylinid beetles *(Myrmedonia)* and a host of other insects. The number of species of insects alone that are known to live in ant and termite colonies is now in the thousands, and many of them have evolved so that they resemble their hosts not only in colour but also in shape and behaviour.

Wasmannian mimicry, in which unwanted guests or 'inquilines' resemble their hosts, provides an example of natural deception in which the host is both the model and the selective agent. The similarity of wasmannian mimics to ants and termites covers not only size, coloration and body structure, but also behaviour. In some cases, mimics never emerge from the dark inner chambers of the ants' nests and do not look anything like their models. Nevertheless, they manage to deceive the ants with which they live, not through the optical sense, but by smell and tactile responses. By associating so closely with their prey, predators themselves become modified and come to resemble the prey in more ways than one.

Some termite guests, likewise, show morphological similarities to their hosts. In the case of staphylinid beetles *(Coatonachthodes)*, for example, the membranous abdomen is bent anteriorly, covering the head and thorax of the beetle completely. The tip of the abdomen then looks like the head of a termite, and membranous abdominal processes resemble its legs. The mimicry must be based on tactile and chemotactic senses, however, since the beetles and their models are in continuous darkness. Such ant-like species have sometimes been considered as batesian mimics, but it is difficult to imagine a vertebrate predator spending hours beside an ant column, picking out the occasional inquiline. True wasmannian mimics of ants are seldom exposed to vertebrate predators.

Although Wasmann clearly distinguished between ant-like inquilines living inside ant colonies, and batesian mimics outside, the vision of insects is so poor that it does not seem possible that selection by their hosts can always account for the extreme ant-like appearance of many inquilines. Moreover, staphylinid beetles and other ant guests often defend themselves by powerful repugnatorial discharges. The selecting agents, then, are probably lizards, birds and other vertebrate insectivores that hunt by sight. The types of mimicry shown by ant guests must, therefore, in some cases be batesian or müllerian rather than wasmannian.

Some of the insects which are associated with ants are welcome guests because they provide their hosts with food. As mentioned earlier, aphids, in particular, excrete large quantities of a sweet fluid appropriately known as 'honeydew', which is collected by ants. It is believed that, in the first instance, ants mistake the posterior of an aphid for the head of another ant, solicit food, and receive a mouthful of honey dew as a reward. In return for this honeydew, ants shepherd their aphids and protect them from predatory enemies.

In one extraordinary instance, a predacious insect larva which feeds on aphids copes with their attendant ants by masquerading as an aphid. This larva, a member of the Neuroptera or lacewings, lives among colonies of the American woolly alder-aphid *(Prociphilus)* on which it feeds: it is not found with any other species of aphid. *Prociphilus* derives its woolly appearance from the fluffy investiture of brilliant white wax that covers its body, making it extremely conspicuous against the dark branches of the alder bushes on which it lives. Woolly alder-aphids are guarded by ants of at least three species, which consume the honeydew they produce. The predatory behaviour of their enemy, the lace-wing larva *(Chrysopa)*, is in no way unusual. It pierces aphids with its hollow, sickle-shaped mandibles, sucking out their contents and discarding the shrivelled remains. What is remarkable is the fact that the larva covers its back with waxy material taken from its prey. This is loosely attached to the long bristles on its back, and results in astonishingly effective mimicry. The lace-wing larvae not only match their prey in size and shape, but move very little, even when not feeding, so that they are extremely hard to detect. The waxy material can easily be removed from the backs of the *Chrysopa* larvae with forceps or a paint brush. When this is done, the denuded larvae are attacked by ants and dragged away. As soon as they have fixed more wax on to their backs, however, they are virtually ignored. Another instance of aggressive mimicry is provided by the South American scale-eating fish *(Probolodus)*, which normally shoals with its fish model *(Astyanax)*. Although the latter does not apparently recognize it as a predator, *Probolodus* does, nevertheless, attack its models, biting off their scales and swallowing them.

An interesting example of ploy and counterploy in predator-prey interactions is afforded by orb-weaving spiders and bombardier-beetles *(Brachinus)* in America. Bombardier-beetles are either captured or lost, depending on the tactics of the spider's attack. *Nephila* grasps a beetle directly and attempts to bite it immediately, but is repelled by the beetle's defensive spray. Copious fluid oozes from its mouth and the spider cleans its jaws for several minutes before it recovers. During this time, the beetle is able to escape from the spider's web, working its way downwards along the radial threads of the orb until it reaches the lower margin and drops to freedom. In contrast, *Argiope* first imprisons the beetle by wrapping it delicately in silk, without causing it to spray. When the spider then proceeds to bite, the silk wrapping protects it against the full effects of the spray. Clearly the wrapping procedure has evolved not only as a defence against injury from larger, struggling prey, but also as a means of coping with insects that are chemically protected. No doubt the direct attack of *Nephila* is more effective against some other kinds of insects prey.

The reactions of an animal to a predator may vary according to circumstances. In the case of wading birds, pure escape is the most usual reaction to man – outside the breeding season. Concealing movements, too, are often evoked. Little ringed plover chicks, for example, when very small and inexperienced, crouch flat immediately on hearing their parents' alarm calls. They remain motionless until they hear one of their parents notifying them that the danger has passed. Larger chicks, on the other hand, do not immediately conceal themselves – the adult's call only puts them on their guard.

In their own territory, breeding birds especially those with eggs or young, rarely show escape behaviour. Their escape reactions appear to be inhibited to some extent. They may run away from an intruder in a posture of concealment, with legs bent, neck contracted, and body horizontal. Then they stop, bob up and down – which may serve to deflect attack from the eggs or young – and give alarm cries. The fact that plovers do not fly right away, but move around the intruder, often giving threat calls as well as alarm cries, shows a tendency towards aggression: they also indulge in injury feigning. Finally, parent waders may remove their young from danger, holding the chicks close to their body between the thighs or, more frequently, inducing them to follow on foot.

The bond between mother and offspring is essential to the survival of young ungulates. By removing the foetal membranes, faeces and other smelly substances, mother antelopes reduce the probability of their young being detected by enemies. Hyenas have been observed to pass within 2–3 metres of Thomson's gazelle fawns without noticing them. Because the mother also defends her babies, predators such as hyenas and hunting dogs may have to band together if they are to make a successful attack. The attempts of predators to separate an individual antelope from the herd is often counteracted by the herd bunching. Similarly, clusters of aphids form tight groups upon being threatened by an ichneumon wasp. In this way, the parasite is prevented from penetrating into the middle of the cluster, and only the individuals at the edge are vulnerable. Animals as dissimilar as sea-slugs, fishes, tadpoles, birds, mongooses and antelopes will bunch together upon sensing a predator, or on being attacked.

Most animals tend to become exceptionally shy in regions where they are often pursued. They may even change their normal time of activity. Young Nile crocodiles become nocturnal in parts of Natal where they are repeatedly attacked by African fish-eagles during the day. Elsewhere, they remain day-active. Consequently, fish-eagles have to find alternative sources of food. No doubt, from time to time, a prey species will become almost exterminated because it is unable to defend itself adequately against its enemies. Conversely, predators may be unable to capture sufficient prey in order to survive. Between 1961 and 1966, lions reduced a herd of wildebeest in Nairobi National Park from 1,780 to 253 individuals. By 1967, however, the wildebeest population had become so low that the lions turned to alternative prey, and the wildebeest numbers began to increase again. It is unlikely that, in natural conditions, a predatory species other than Man ever completely exterminates its prey.

A predator may exploit several prey species while, conversely, a prey species may be exploited by several different predators. In Arctic regions, where the number of species of both predator and prey is low, population sizes tend to fluctuate widely from year to year. Correlated with cyclical variations in the numbers of hares, lemmings and, further south, of voles, there are cycles in the numbers of their predators such as wolves, foxes and owls, which increase as more food becomes available. When, despite increased predation, the numbers of herbivores reach a peak, mass emigrations take place, during which most of the emigrants perish and populations crash. The mass migrations of lemmings have become legendary, but they are not suicidal – the popular misconception. In contrast to conditions in Arctic regions, the enormous diversity of plant and animal species in tropical rainforest ensures that numbers do not fluctuate too widely – unless, of course, the inherent stability of the ecosystem has been destroyed by human activity.

A predator or a parasite that causes a great decrease in the numbers of its prey or host is also elminating its own food supply. This is particularly evident when the host or prey consists of a single species; but it is less apparent when many different species are simultaneously exploited. More efficient adaptation, however, is attained through specialization, and the two opposing tendencies lead towards a balanced compromise – for natural selection must favour adaptations that tend to bring opposing systems into equilibrium. Predation is, therefore, not usually harmful to the prey species. A state of equilibrium may even develop that is beneficial both to prey and to predator. Excess of an herbivorous species can result in overgrazing and starvation if the population outgrows its food supply – as in the case of lemmings. This does not happen when its numbers are controlled by predation. Even ruthless killers, such as hunting dogs, are frequently useful in causing new blood to be introduced into herds of antelope. For example, a group of impala may be dominated and served by a single male for a number of years, with consequent inbreeding, until attack by the dogs causes such havoc that the herd is split up and joined by other males so that the tyrant loses control.

When different kinds of animal exploit similar habitats and are affected by similar ecological factors, the environmental influence upon evolution becomes very clear. The phenomenon of parallel evolution occurs, and similar adaptational changes take place in comparatively unrelated species. For example, the mammals that have become adjusted to a diet of ants and termites all have cylindrical tongues and a reduction in the number of teeth. Examples appear in six different orders of mammals – the spiny ant-eater, a monotreme related to the platypus; the banded ant-eater, a marsupial related to the kangaroo; the aardvark (Tubulidentata); the pangolin (Pholidota); the aardwolf *(Proteles)*, a carnivore; and the New World ant-eater (Xenarthra). All except the banded ant-eater and aardwolf have specialized fossorial feet for digging into the nests of their prey.

Our old friend the aardwolf has a diet restricted mainly to termites of the genus *Trinervitermes*. He does not dig for food, but detects parties of worker termites foraging in the open at night – probably mainly by sound but also perhaps by the sense of smell. At any rate, the aardwolf always turns upwind when it comes upon termites. Having located them, the aardwolf rapidly licks them up with its broad tongue, pushing the front part of the snout, which is black and almost hairless, deep into the vegetation. When too many soldier termites appear, discharging their chemical defensive secretions, the aardwolf goes off to find another party of defenceless workers! During the rains, the aardwolf's diet increases in variety, probably because *Trinervitermes* workers disappear underground then, while army-ants and other termites are more readily available. Although the aardwolf is an exception, in general large predators tend to feed on larger prey, while smaller predatory species take smaller prey – unless they combine, like wolves and hyenas, to hunt in a pack. The number of predators is related to the size of the prey population so that a balance is achieved in nature. In years when lemmings are plentiful, snowy owls tend to hatch two broods of young, while many carnivores show a reduced rate of reproduction when food is scarce. Starvation is the ultimate factor controlling animal populations, although it is probably seldom invoked. There are other ways, too, in which the numbers of predators are adjusted to those of their herbivorous prey. Carnivores tend to have slower reproductive cycles, range further, and are less strictly confined to one habitat than are herbivores.

In Serengeti, the greatest competition between predators is for the smaller species of prey, the gazelles. If these are scarce, lions and hyenas can always turn to alternative, larger food such as wildebeest or buffalo but, for jackals and cheetahs which are less powerful, this is not possible. On the other hand, a Thomson's gazelle provides comparatively small reward in terms of food per kill, so a large predator must either have a high rate of success when hunting, or else use a method which takes little time and energy. Because of the competition between predators for the available prey, each species has evolved specializations. Lions and hyenas are specialists at tackling large and formidable prey. The lion stalks buffalo, wildebeest, zebras and gazelles: the hyena courses its prey and is also a highly efficient scavenger. Leopards, cheetahs, wild dogs and jackals specialize on smaller prey. The leopard stalks a variety of antelope species while the other three are coursers and chase their prey. Jackals are mainly scavengers, but can also kill adult gazelles and the young of larger herbivores, while cheetahs are extremely fast and rely for over 90 per cent of their kills on gazelles. Because of their social habits, hunting dogs are able to tackle larger game such as wildebeest and zebras, but their success rate when hunting the latter is very low.

If, therefore, gazelles were to become scarce, cheetahs would probably suffer most. On the other hand, as hyenas can only capture the oldest or most diseased of the gazelle herds, and because cheetahs are more highly specialized at capturing gazelles and would still catch those that could outrun a hyena, the latter would have to find alternative prey. Predator-prey systems have evolved a balance so that they can survive in equilibrium even though considerable fluctuations in numbers may be involved.

There are, however, odd occasions when the delicate balance between numbers of predators and of prey may be upset by carnivores killing a great many more prey animals than they can possibly eat at any one time. Surplus killing can be spectacular, as when a fox gets into a hen-house, and it also occurs under completely natural conditions. Instances have been cited of foxes raiding colonies of black-headed gulls. Up to 230 birds were killed by each fox in one case when the gulls were nesting on sand dunes in Cumberland, but less than 3 per cent of the victims were eaten. Although foxes ate only a small percentage of the gulls they killed, analysis of their faeces showed that gulls nevertheless formed 26 per cent of their diet, the rest being chiefly rabbits and Sandwich terns – which occupied small nesting areas within the gull colony. Tracks in the sand near the carcasses showed that the birds had made little or no attempt to escape: the foxes' forays had occurred only on very dark nights. Poor visibility also plays a part in the mass slaughter of Thomson's gazelle by spotted hyenas in Serengeti. On one occasion, 82 gazelles were found dead, although very few had been eaten. Hyena tracks near the bodies gave no indication of a chase, although, under normal circumstances, a gazelle may be chased for up to 10 km.

Surplus slaughter on this scale seems not only to be wasteful, but to be against the killer's own interests. As well as depleting potential food supplies, the predator wastes energy and risks physical injury unnecessarily. On the other hand, of course, useful practice is gained in the art of killing. Unnecessary slaughter has been observed in other animals. Lions, for example, continue killing wildebeest even when they are gorged with meat. Lions that get into enclosures have killed large numbers of captive cattle, ostriches or donkeys without eating them. Narwhals, too, have been slaughtered in large numbers when trapped by ice in a small pool which made them easy prey for a polar bear. There is a report of 21 narwhals killed at one time; but it was sufficiently cold for the animals to be preserved for later consumption.

A few carnivores do deliberately hoard food. The leopard, for instance, makes a treetop larder to which it later returns. But the species that hoard are in a minority, and are restricted chiefly to members of the cat family. Moreover, most of the carnivores that store food do so haphazardly, often not returning to the cache until the meat is so decayed that it is inedible. A carnivorous animal, then, might be expected to kill only when hungry – after feeding, it no longer wants to search or hunt for food. But, if it stumbles upon easy prey, the normal behaviour pattern for killing is stimulated – there is no 'feed-back' to prevent the sated predator from continuing to kill easy prey. Unnecessary killing results as much from failure of the prey to make an escape from the predator as from the instinct of the latter to kill. On a pitch-black night it would be very dangerous for any prey animal to move at all, so it normally keeps still until conditions for flight are more favourable. Fortunately for the prey, suitable conditions for flight usually occur before a predator arrives on the scene!

From an ecological point of view, the rarity of the coincidence of the various factors necessary to produce a large mass-kill indicates that the consequences of surplus killing are usually small. Even when mass-killings do occur, it is by no means certain that this immediately deprives the carnivore of part of its resources – the individuals involved are not necessarily the ones that would normally be preyed on by the particular species of carnivore. Moreover, the carnivore population is not necessarily limited by food supply in a way that surplus killing could upset. Surplus killing should be viewed not as the consequence of mere maladaptations, but as a necessary and relatively small disadvantage accompanying behavioural compromises to meet conflicting environmental demands on both carnivores and their prey. To prevent its occurrence, carnivores would have to possess a special inhibition of their hunting and killing behaviour, not after feeding to satiation but after the receipt of information that further carcasses could not be utilized. Surplus killing only takes place when relatively defenceless prey animals, which normally respond to their enemies by flight, are prevented from so doing by fog, exceptional darkness or enclosing fences.

All the intricate defence mechanisms that have been described in this book are in continuous operation throughout the world.

Tooth & Claw

Some particular examples may disappear as their possessors are supplanted or become extinct but, so long as animal life persists, the principles of camouflage and disguise, of evasion and deterrence will be invoked – for they are part of natural selection, the driving force of evolution. In this sense, the world we have been examining is a static one. The principles of defence have, no doubt, been exploited to the full for many hundreds of million years, and will continue for hundreds of millions more. We read most about hunting, fighting and aggression but, to use the words of John Ruskin, 'there was always more in the world than men could see, walked they ever so slowly'. A brief pause, therefore, to consider another and less well-known of the many aspects of nature may serve, not only to refresh our minds, but also to revive interest in the wonderful things we sometimes fail to notice in the hustle of everyday life. We are then obliged to confess what we should long ago have known, that the really precious things of life are thought and sight, not merely movement and change for their own sakes. 'It does a bullet no good to go fast; and a man, if he be truly a man, no harm to go slow; for his glory is not all in going, but in being.'

Bibliography

No attempt has been made at complete documentation. Sufficient references are supplied, however, for the interested reader to follow up the various topics discussed.

Introduction

Lull, R. S. (1940). *Organic Evolution* (rev. ed.) New York: Macmillan. 743 pp.

1 Life in a Hole

Bennett, M. F. (1974). *Living Clocks in the Animal World*. Springfield, Ill.: Thomas. xiii+221 pp.

Cloudsley-Thompson, J. L. (1961). *Rhythmic Activity in Animal Physiology and Behaviour*. New York and London: Academic Press. vii+236 pp.

Cloudlsey-Thompson, J. L. (1980). *Biological Clocks. Their Functions in Nature*. London: Weidenfeld and Nicolson. viii+144 pp.

Cloudsley-Thompson, J. L. and Chadwick, M. J. (1964). *Life in Deserts*. London: Foulis. xvi+218 pp.

Edmunds, M. (1974). *Defence in Animals*. Harlow, Essex: Longman. xvii+357 pp.

Lawrence, R. F. (1953). *The Biology of the Cryptic Fauna of Forests*. Cape Town: Balkema. 408 pp.

Russell, F. S. & Yonge, C. M. (1975). *The Seas* (4th ed.) London: Warne. 283 pp.

Savory, T. (1971). *Biology of the Cryptozoa*. Watford, Herts: Merrow. vii+48 pp.

Trueman, E. R. (1975). *The Locomotion of Soft-Bodied Animals*. London: Arnold. viii+200 pp.

Bibliography

2   Camouflage

Chapman, G. (1976). Reflections on transparency. pp. 491–9 *in* Mackie, G. O. (ed.) *Coelenterate Ecology and Behaviour*. New York: Plenum Press. xiii+744 pp.

desert animals. *Journal of Arid Environments*, **2**: 95–104.

Cott, H. B. (1940). *Adaptive Coloration in Animals*. London: Methuen. xxxii+508 pp.

Hamilton, W. J., III (1973). *Life's Color Code*. New York: McGraw-Hill. x+238 pp.

Stephenson, E. M. (1946) *Animal Camouflage*. Harmondsworth, Middx: Penguin. 160 pp.

3   Disguise

Breder, C. M. (1949). On the behaviour of young *Lobotes surinamensis*. *Copeia*, **1949**: 237–42.

Cott, H. B. (1940). *Adaptive Coloration in Animals*. London: Methuen. xxxii+508 pp.

Edmunds, M. (1974). *Defence in Animals*. Harlow, Essex: Longman. xvii+357 pp.

Fitzpatrick, T. W. (1957). *Insect Life in the Tropics*. London: Longmans, Green. xiv+311 pp.

Hinton, H. E. (1955). Protective devices of endopterygote pupae. *Transactions of the Society for British Entomology*, **12**: 49–92.

Hinton, H. E. (1974). Lycaenid pupae that mimic anthropoid heads. *Journal of Entomology* (A) **49**: 65–9.

Hinton, H. E. (1973). Natural deception. pp. 97–159 *in* Gregory, R. L. and Gombrich, E. H. *Illusion in Nature and Art*. London: Duckworth. 288 pp.

Poulton, E. B. (1980). *The Colours of Animals. Their Meaning and Use*. London: Kegan Paul, Trench, Trubner. xv+360 pp.

Wickler, W. (1968). *Mimicry in Plants and Animals*. London: Weidenfeld and Nicolson. 255 pp.

4   Armour

Boulière, F. (1955). *The Natural History of Mammals*. London: Harrap. xxi+363+xi pp.

Colbert, E. H. (1962). *Dinosaurs. Their Discovery and Their World*. London: Hutchinson. 288 pp.

Edmunds, M. (1966). Protective mechanisms in the Eolidacea (Mollusca Nudibranchia). *Journal of the Linnean Society (Zoology)*, **46**: 27–71.

Packard, A. (1972). Cephalopods and fish: the limits of convergence. *Biological Reviews*, **47**: 241–307.
Thorson, G. (1971). *Life in the Sea.* (Transl. M. C. Meilgaard and A. Laurie). London: Weidenfeld and Nicolson. 256 pp.
Yonge, C. M. (1949). *The Sea Shore.* London: Collins. xvi+311 pp.

5 Barbs and Spines

Boulière, F. (1955). *The Natural History of Mammals.* London: Harrap. xxi+363+xi pp.
Bücherl, W. and Buckley, E. E. (eds.) (1971). *Venomous Animals and Their Venoms.* Vol. III. *Venomous Invertebrates.* New York and London: Academic press, xxii+537 pp.
Eisner, T. (1970). Chemical defence against predation in arthropods. pp. 157–215 *in* Sondheimer, E. and Simeone, J. B. (eds.) *Chemical Ecology.* New York and London: Academic Press. xv+366 pp.
Evans, H. M. (1943). *Sting-fish and Seafarer.* London: Faber and Faber. 180 pp.
Halbach, U. (1971). Zum Adaptivwert der zyklomorphen Dornenbildung von *Brachionus calyciflorus* Pallas (Rotatoria). 1. Räuber-Beute-Beziehung in Kurzzeit-Versuchen. *Oecologia (Berlin)*, **6**: 267–88.
Hinton, H. E. (1946). The 'gin-traps' of some beetle pupae: a protective device which appears to be unknown. *Transactions of the Royal Entomological Society of London*, **97**: 473–96.
Hoogland, R., Morris, D. and Tinbergen, N. (1957). The spines of sticklebacks *(Gasterosteus* and *Pygosteus)* as means of defence against predators. *(Perca* and *Esox). Behaviour*, **10**: 205–36.
Morris, D. (1958). The reproductive behaviour of the ten-spined stickleback *(Pygosteus pungitius* L). *Behaviour* (Suppl.), **6**: 1–154.

6 Chemical Defences

Boulière, F. (1955). *The Natural History of Mammals.* London: Harrap. xxi+363+xi pp.
Cloudsley-Thompson, J. L. (1968). *Spiders, Scorpions, Centipedes and Mites* (2nd ed.) Oxford, London and New York: Pergamon. xv+278 pp.
Cott, H. B. (1946). The edibility of birds. *Proceedings of the Zoological Society of London*, **116**: 371–524.
Cott, H. B. (1954). The palatability of the eggs of birds. *Proceedings of the Zoological Society of London*, **124**: 335–463.

Bibliography

Crawford, C. S. and Cloudsley-Thompson, J. L. (1971). Concealment behaviour of nymphs of *Blaberus giganteus* L. (Dictyoptera: Blattaria) in relation to their ecology. *Revista de Biología Tropical*, **18**: 53–61.

Edmunds, M. (1971). *Defence in Animals*. Harlow, Essex: Longman. xvii+357 pp.

Eisner, T. (1970). Chemical defence against predation in arthropods. pp. 157–215 *in* Sondheimer, E. and Simeone, J. B. (eds.) *Chemical Ecology*. New York and London: Academic Press. xv+336 pp.

Eisner, T., Kriston, I. and Aneshansley, D. J. (1976). Defensive behaviour of a termite *(Nasutitermes exitiosus)*. *Behavioural Ecology and Sociobiology*, **1**: 83–125.

Kalmijn, A. J. (1971). The electric sense of sharks and rays. *Journal of Experimental Biology*, **55**: 371–83.

Kirk, V. M. and Dupraz, B. J. (1972). Discharge by a female ground beetle *Pterostichus lucublandus* (Coleoptera, Carabidae), used as a defence against males. *Annals of the Entomological Society of America*, **65**: 513.

7 Venoms

Beard, R. L. (1963). Insect toxins and venoms. *Annual Review of Entomology*, **8**: 1–18.

Bellairs, A. (1969). *The Life of Reptiles*. Vol. I. London: Weidenfeld and Nicolson. xii+282 pp.

Bücherl, W., Buckley, E. E. and Deulofeu, V. (eds.) (1967). *Venomous Animals and Their Venoms*. Vol. I. *Venomous Vertebrates*. New York and London: Academic Press. xxii+707 pp.

Bücherl, W. and Buckley, E. E. (eds.) (1971). *Venomous Animals and Their Venoms*. Vol. II. *Venomous Vertebrates*. New York and London: Academic Press. xxiv+687 pp.

Bücherl, W. and Buckley, E. E. (eds.) (1971) *Venomous Animals and Their Venoms*. Vol. III. *Venomous Invertebrates*. New York and London: Academic Press. xii+537 pp.

Cloudsley-Thompson, J. L. (1965). The scorpion. *Science Journal*, **1** (4): 35–41.

Cloudsley-Thompson, J. L. (1976). *Insects and History*. London: Weidenfeld and Nicolson. 242 pp.

Cloudsley-Thompson, J. L. (1978) Biological clocks in Arachnida. *Bulletin of the British Arachnological Society*, **4**: 184–91.

Ditmars, R. L. (1931). *Snakes of the World*. New York: Macmillan. xi+207 pp.

Bibliography

Hinton, H. E. and Dunn, A. M. S. (1967). *Mongooses: Their Natural History and Behaviour*. Edinburgh and London: Oliver and Boyd. vi+144 pp.

Minton, S. A. Jr. (1968). Venoms of desert animals. pp. 487–516 *in* Brown, G. W. Jr. (ed.) *Desert Biology*. Vol. I. New York and London: Academic Press. xvii+635 pp.

Pavan, M. and Dazzini, M. V. (1971). Toxicology and pharmacology – Arthropoda. pp. 365–409 *in* Florkin, M. and Scheer, B. T. (eds.) *Chemical Zoology*. Vol. VI. New York and London: Academic Press. xix+484 pp.

Phisalix, M. (1922). *Animaux Venimeux et Venins*. Paris: Masson. Tome I xxv+656 pp. Tome II. xii+864 pp.

Russell, F. E. and Brodie, A. F. (1974). Venoms of reptiles. pp. 449–78 *in* Florkin, M. and Scheer, B. T. (eds.) *Chemical Zoology*. Vol. IX. New York and London: Academic Press.

8 Warning and Threat

Cott, H. B. (1940). *Adaptive Coloration in Animals*. London: Methuen. xxxii+508 pp.

Dumortier, B. (1963). Morphology of sound emission apparatus in Arthropoda. pp. 277–345. Ethological and physiological study of sound emissions in Arthropoda. pp. 583–654. *In* Busnel, R.-G. (ed.) *Acoustic Behaviour of Animals*. Amsterdam, London, New York: Elsevier. 933 pp.

Edmunds, M. (1974). *Defence in Animals*. Harlow, Essex: Longman. xvii+357 pp.

Poulton, E. B. (1890). *The Colours of Animals. Their Meaning and Use*. London: Kegan Paul, Trench, Trubner. xiii+360 pp.

Wickler, W. (1968). *Mimicry in Plants and Animals*. London: Weidenfeld and Nicolson. 255 pp.

9 Mimicry

Baker, R. P. and Parker, G. A. (1979). The evolution of bird coloration. *Philosophical Transactions of the Royal Society of London*, B **287**: 63–130.

Brower, L. P. (1971). Prey coloration and predatory behaviour. pp. 66–76 *in* Kramer, A. (ed.) *Topics in Animal Behaviour*. New York and London: Harper & Row. x+77 pp.

Brower, L. P and Brower, J. V. Z. (1961). Investigations into mimicry. *Natural History*, **71** (4): 8–19.

Bibliography

Cott, H. B. (1940). *Adaptive Coloration in Animals*. London: Methuen. xxxii+508 pp.

Edmunds, M. (1974). *Defence in Animals*. Harlow, Essex: Longman. xvii+357 pp.

Grobman, A. B. (1978). An alternative solution to the coral snake mimic problem, (Reptilia, Serpentes, Elapidae). *Journal of Herpetology*, **12**: 1–11.

Hespenheide, H. A. (1973). A novel mimicry complex: beetles and flies. *Journal of Entomology* (A), **48**: 49–56.

Hingston, R. W. G. (1927). Field observations on spider mimics. *Proceedings of the Zoological Society of London*, **1927**: 841–58.

Holm, E. and Kirsten, J. F. (1979). Pre-adaptation and speed mimicry among Namib Desert scarabaeids with orange elytra. *Journal of Arid Environments*, **2**: 263–71.

Lane, C. and Rothschild, M. (1965). A case of Müllerian mimicry of sound. *Proceedings of the Royal Entomological Society of London* (A) **40**: 156–8.

Mertens, R. (1966). Das Problem der Mimikry bei Korallenschlangen. *Zoologische Jahrbücher Abteilung für Systematik, Geographie und Biologie der Tiere*, **84**: 541–76.

Rettenmeyer, C. W. (1970). Insect mimicry. *Annual Reviews of Entomology*, **15**: 43–74.

Thiele, H.-U. (1977). *Carabid Beetles in Their Environments*. Berlin and New York: Springer-Verlag. xvii+369 pp.

Vane-Wright, R. I. (1976). A unified classification of mimetic resemblances. *Biological Journal of the Linnean Society*, **8**: 25–56.

Wickler, W. (1968). *Mimicry in Plants and Animals*. London: Weidenfeld and Nicolson. 255 pp.

10 Bluff, Death Feigning and Deflection of Attack

Cloudsley-Thompson, J. L. (1961). A new sound-producing mechanism in centipedes. *Entomologists' Monthly Magazine*, **96**: 110–13.

Cott, H. B. (1940). *Adaptive Coloration in Animals*. London: Methuen. xxxii+508 pp.

Eberhard, W. G. (1973). Stabilimenta on the webs of *Uloborus diversus* (Araneae: Uloboridae) and other spiders. *Journal of Zoology, London*, **171**: 367–84.

Edmunds, M. (1974). *Defence in Animals*. Harlow, Essex: Longman. xvii+357 pp.

Ewer, R. F. (1968). *Ethology of Mammals*. London: Logos Press. xiv+418 pp.

Hinton, H. E. (1973). Natural deception. pp. 96–159 *in* Gregory, R. L. and Gombrich, E. H. (eds.) *Illusion in Nature and Art*. London: Duckworth, 288 pp.

Jourdain, F. C. R. (1936) The so-called 'injury feigning' in birds. *Oologists' Record, London,* **16**: 25–37; 62–70.

Robinson, M. H., Abele, L. G. and Robinson, B. (1970). Attack autotomy: defence against predators. *Science,* **169**: 300–1.

Robinson, M. H. and Robinson, B. (1970). The stabilimentum of the orb web spider, *Argiope argentata*: an improbable defence against predators. *Canadian Entomologist,* **102**: 641–55.

Stasek, C. R. (1967). Autotomy in the Mollusca. *Occasional Papers of the California Academy of Science,* no. **61**: 1–44.

Vitt, L. J., Congdon, J. D. and Dickson, N. A. (1977). Adaptive strategies and energetics of tail autotomy in lizards. *Ecology,* **58**: 326–37.

11 Withdrawal and Flight

Ewer, R. F. (1968). *Ethology of Mammals*. London: Logos Press. xiv.+418 pp.

Gray, J. (1968). *Animal Locomotion*. London: Weidenfeld and Nicolson. xi+479.

Harkness, R. D. (1977). Locomotion. pp. 134–62 *in* Bligh, J., Cloudsley-Thompson, J. L. and Macdonald, A. G. (eds.) *Environmental Physiology of Animals*. Oxford: Blackwell. vii+456 pp.

Humphries, D. A. and Driver, P. M. (1970). Protean defence by prey animals. *Oecologia (Berlin),* **5**: 285–302.

Leuthold, W. (1977). *African Ungulates. A Comparative Review of Their Ethology and Behavioral Ecology*. Berlin, Heidelberg, New York: Springer-Verlag. vii+307 pp.

Sales, G. and Pye, D. (1974). *Ultrasonic Communication by Animals*. London: Chapman and Hall. xi+281 pp.

12 Horns, Teeth and Claws

Geist, V. (1966). The evolution of horn-like organs. *Behaviour,* **27**: 175–214.

Hopson, J. A. (1977). Relative brain size and behaviour in archosaurian reptiles. *Annual Review of Ecology and Systematics,* **8**: 429–48.

Scullard, H. H. (1974). *The Elephant in the Greek and Roman World*. London: Thames and Hudson. 288 pp.

13 Co-operation

Alexander, R. D. (1974). The evolution of social behaviour. *Annual Review of Ecology and Systematics*, **5**: 325–83.

Cloudsley-Thompson, J. L. (1961). *Rhythmic Activity in Animal Physiology and Behaviour.* New York and London: Academic Press. vii+236 pp.

Cott, H. B. (1940). *Adaptive Coloration in Animals.* London: Methuen. xxxii+508 pp.

Couch, J. N. (1938). *The Genus Septobasidium.* Chapel Hill: University of North Carolina Press. ix+480 pp.

Dawkins, R. (1976). *The Selfish Gene.* Oxford: Oxford University Press. xii+224 pp.

Estes, R. D. (1976). The significance of breeding synchrony in the wildebeest. *East African Wildlife Journal*, **14**: 135–52.

Frost, S. W. (1942). *General Entomology.* New York and London: McGraw-Hill. x+524 pp.

Gillett, S. D., Hogarth, P. J. and Noble, F. E. J. (1979). The response of predators to varying densities of *gregaria* locust nymphs. *Animal Behaviour*, **27**: 592–6.

Harborne, J. B. (1977) *Introduction to Ecological Biochemistry.* London: Academic Press. ix+243 pp.

Harvey, P. H. and Greenwood, P. J. (1978). Anti-predator defence strategies – some evolutionary problems. pp. 129–51 *in* Krebs, J. J. and Davies, N. B. (eds.) *Behavioural Ecology. An Evolutionary Approach.* Oxford: Blackwell Scientific Publications. xi+494 pp.

Humphries, D. A. and Driver, P. M. (1970). Protean defence by prey animals. *Oecologia (Berlin)*, **5**: 285–302.

Leuthold, W. (1977). *African Ungulates. A Comparative Review of Their Ethology and Behavioral Ecology.* Berlin, Heidelberg, New York: Springer-Verlag. xiii+307 pp.

Rothschild, M. (1972). Some observations on the relationship between plants, toxic insects and birds. pp. 1–12 *in* Harborne, J. B. (ed.) *Phytochemical Ecology.* London and New York: Academic Press. xiv+272 pp.

Smith, N. G. (1968). The advantages of being parasitized. *Nature, London*, **219**: 690–4.

Ward, P. (1972). The functional significance of mass drinking flights by sandgrouse: Pteroclididae. *Ibis*, **114**: 533–6.

Way, M. J. (1963). Mutualism between ants and honey-dew-producing Homoptera. *Annual Review of Entomology*, **8**: 307–44.

Wilson, E. O. (1976). the organization of colony defence in the ant *Pheidole dentata* Mayr (Hymenoptera: Formicidae). *Behavioral Ecol-*

ogy and Sociobiology, **1**: 63–81.
Wynne-Edwards, V. C. (1962). *Animal Dispersion in Relation to Social Behaviour.* Edinburgh and London: Oliver and Boyd. xi+653 pp.

14 Predator-Prey Interactions

Cloudsley-Thompson, J. L. (1965). *Animal Conflict and Adaptation.* London: Foulis. xi+160 pp.

Crane, J. (1952). A comparative study of innate defensive behaviour in Trinidad mantids (Orthoptera, Mantoidea). *Zoologica,* **37**: 259–93.

Curio, E. (1976). *The Ethology of Predation.* Berlin, Heidelberg, New York: Springer-Verlag. x+250 pp.

Donisthorpe, H. St. J. K. (1927). *The Guests of British Ants. Their Habits and Life-Histories.* London: Routledge. xxiii+244 pp.

Edmunds, M. (1972). Defensive behaviour in Ghanaian praying mantids. *Zoological Journal of the Linnean Society,* **51**: 1–32.

Eisner, T., Hicks, K., Eisner, M. and Robson, D. S. (1978). 'Wolf-in-sheep's clothing' strategy of a predaceous insect larva. *Science,* **199**: 790–4.

Ewer, R. F. (1963). The behaviour of the meerkat, *Suricata suricatta* (Schreber). *Zeitschrift für Tierpsychologie,* **20**: 570–607.

Harvey, P. H. and Greenwood, P. J. (1978). Anti-predator defence strategies: some evolutionary problems. pp. 129–151 *in* Krebs, J. R. and Davies, N. B. (eds.) *Behavioural Ecology. An Evolutionary Approach.* Oxford: Blackwell. xi+494 pp.

Kruuk, H. (1972). Surplus killing by carnivores. *Journal of Zoology, London,* **166**: 233–44.

Murphy, J. B., Lamoreaux, W. E. and Carpenter, C. C. (1978). Threatening behaviour in the angle-headed dragon, *Goniocephalus dilophus* (Reptilia, Lacertilia, Agamidae). *Journal of Herpetology,* **12**: 455–60.

Rettenmeyer, C. W. (1970). Insect mimicry. *Annual Review of Entomology,* **15**: 43–74.

Robinson, M. H. (1969). The defensive behaviour of some orthopteroid insects from Panama. *Transactions of the Royal Entomological Society of London,* **121**: 281–303.

Simmons, K. E. L. (1955). The nature of the predator-reactions of waders towards humans; with special reference to the role of the aggressive-, escape- and brooding-drives. *Behaviour,* **8**: 130–73.

Wasmann, E. (1925). Die Ameisenmimikry, ein exakter Beitrag zum Mimikryproblem und zur Theorie der Anpassung. *Abhandlungen zur theoretischen Biologie,* **19**: xii+164 pp.

Index

Generic and specific names are in italics. Numerals in bold type denote pages on which there are illustrations. Where only English names have been given in the text, further taxonomic details are provided wherever necessary to make the book more precise for the use of students and professional zoologists.

Aardvark *(Orycteropus afer)*, 228
aardwolf *(Proteles cristata)*, 76, 145–6, **146**, 161, 228–9
Acanthaspis, 77
Acanthaster, 57 (*see* crown-of-thorns starfish)
Acherontia, 134, **134** (*see* death's head hawk-moth)
Acrocanthosaurus, 200
adder *(Vipera berus)*, 124, 126, 161
adventitious disguise, 53–5, 76, 77, 225
Aelian, 126
aggregation, 130, 201, 205, 211, 226–7
aggressive mimicry, 155, 223, 225
Agkistrodon, 123
agouta *(Solenodon paradoxus)*, 120
alder-aphid *(Procephilus)*, 224–5
Aleyrodidae, 74
alligators (Alligatoridae), 68, 200 (*see* crocodiles)
alligator-bugs (Fulgoridae), 49–51, **50**, 157
alligator-snapper *(Macroclemys temmincki)*, 41
almiqui *(Solenodon cubanus)*, 120
Alphitobius, 74
Altispinax, 200
altruism, 207–8
Ammophila, 176 (*see* sand-wasps)
amphibians, 93, 119
amphisbaenians, 53
Amyciaea, 143, **144**
anal glands, 95–6

anchorites (anachoretes), 17–27, 88, 139, 141, 169, 174
angler-fishes (Lophiidae), 40–1
Anisomorpha, 88 (*see* stick-insects)
Ankylosauria, 68
ants, 24, 67, 74, 75–7, 84, 86, 91, 102, 103–4, 133, 210–11, 212–13, 216–17, 223
ant-eaters, 76, 228–9 (*see under* species)
ant-lions *(Myrmeleon)*, 75–6
ant mimicry, 54, **142**, 142–5
antelopes, 35, 37, 162–3, 181, 182–3, 184, 190–1, 192, 194
Antennarius, 40–1
Antiochus II, 196
antlers, 162, 191, 192, 195–6 (*see* horns)
aphids (Aphididae), 85–6, 212–13, 224–5
Apis, 147 (*see* honey-bees)
apodemes, 170
Apollodorus, 82
aposematic (warning) coloration, 72, 83, 94, 95–6, 127, 128–33, 141, 148, 152–4, 173, 205–6, 217; scents, 138; sounds, 72, 133–9, 153, 162, 178
Arenicola, 169 (*see* lug-worms)
Argiope, 168, 225–6 (*see* orb-web spiders)
A. argentata, 168
Aristotle, 105, 126
Armadillidium, 56 (*see* woodlice)
armadillos (Dasypodidae), 69–70, 71; fairy *(Chlamyphorus)* **175**, 175–6

243

Index

armour, 11, 56-70
Asclepiadaceae, 92
assassin-bugs (Reduviidae), 89, **114**, 114-15
asymmetry, 55
Atrax, 112
autotomy, 63, 186-72; centipedes, 170; crabs, 172; lizards, 171, 223; rodents, 172

Baboons *(Papio)*, 108, 206, 211
bag-worms (Psychidae), 76-7
Balanoglossus, 21
banded ant-eater *(Myrmecobius fasciatus)*, 76, 228
barn-owls *(Tyto)*, 38
Bates, H. W., 140-1
batesian mimicry, 141, 145, 152, 154, 161, 224
batrachotoxins, 93
bats (Chiroptera), 39, 137, 148, 178-80
bees (Apoidea), 102, 133, 135, 146-7, 149, 155, 169, 209-10 *(see* bumble-bees, honey-bees)
bee mimicry, 146-7
beetles (Coleoptera), 24, 54, 66, 72-3, 88-90, 91, 92, 134, 136, 150, 151, 223, 225-6 *(see under* families)
biological clocks, 26-7, 88
birds, coloration, 131, 151; distasteful, 94, 151; hunting methods, 38-9, 50-1, 76, 147-8, 158 *(see under* species)
Bitis, 124 *(see* gaboon-viper, puff-adder)
Blaberus, 88 *(see* cockroaches)
black coloration, 131-2
black-headed gull *(Larus ridibundus)*, 230
black-necked cobra *(Naja melanoleuca)*, 122
bleeding, reflex, 91
blesbok *(Damaliscus albifrons)*, 181
blister-beetles (Meloidae), 91, 92 *(see* cantharidin)
bluejays *(Cyanocitta)*, 92 *(see* jays)
boars, wild *(Sus scrofula)*, 198
Bokhara, Emir of, 115
Bolbonota, 205
bombardier-beetles *(Brachinus)*, 89-90, 134, 255-6
Bombus, 146-7 *(see* bumble-bees)
boomslang *(Dispholidus typhus)*, 125, 160
botflies *(Philornis)*, 217-19
Bothrops, 125
box-tortoise *(Kinixys)*, 67

box-turtles *(Terrapene)*, 66
Brachinus, 255 *(see* bombardier-beetles)
Brachiopoda, 62
brain-corals *(Meandrina)*, 57
brittle-stars (Ophiuroidea), 64, 169
Brower, J. V. Z., 147, 153
Brower, L. P., 147, 155, 205
bugs (Hemiptera) 77, 88, 114-15, 130 *(see* assassin-bugs)
bulldog-ants *(Myrmecia)*, 102
bumble-bees *(Bombus)*, 146-7, 149, 155
Bungarus, 123 *(see* kraits)
burrowing animals, 18-19, 21-7, 175 *(see* anchorites)
burying-beetles (Silphidae), 149
bushbuck *(Tragelaphus scriptus)*, **197**
bushmaster *(Lachesis muta)*, 125
Buthidae, 107 *(see* scorpions)
butterflies, 31, 140; blues (Lycaenidae), 51, 136, 166; colour forms, seasonal, 31; comma *(Polygonia c-album)*, 37; coppers (Lycaenidae), 136; hairstreak *(Thecla)*, 52; Heliconiidae, 140-1; monarch *(Danaus plexippus)*, 92, 151-2 *(see* moths)

Caciques *(Cacicus)*, 211-12
caddis-flies (Trichoptera), 73-4, 76
Caiman, **50**
caimans, 49, **50**, 51 *(see* alligators)
camel *(Camelus bactrianus)*, 162
camel-spiders *(Galeodes)*, 66, 113, 135
camouflage, 28-39, 78, 81, 82-3 *(see* disguise)
Camponotus, **144** *(see* ant mimicry)
cannon bones, 184
cantharidin, 84, 91, 92
Caprimulgus, 165
Carabidae, 88 *(see* ground-beetles)
Carapus, 174 *(see* pearl-fishes)
caribou *(Rangifer tarandus)*, 195-6
Carpenter, G. D. H., 155
carpet-beetles (Dermestidae), 72-3
carpet-viper *(Echis carinatus)*, 124, 137-8
cat-snakes *(Telescopus)*, 121-2
caterpillars, 31, 46, 48-9, 54, 86-7, 129, 130, 157, 161, 180, 205 *(see* larvae); woolly-bear (Lymantriidae), 73
catfish, electric *(Malapturus electricus)*, 97, 99
Catocala, 164
cave animals, 13, 25
centipedes (Chilopoda), 85, 87, **113**, 113-15, 166, 169-70

244

Index

Cepaea, 62
Cephalopoda, 32, 59, 63, 119, 142, 172–3, 178, 188
cerata, 63, 130, 169
Ceratophrys, 159
Ceratosaurus, 200
chameleons (Chamaeleonidae), 159, 183, 189
chamois *(Rupicapra rupicapra)*, 193, 194–5
cheetah *(Acinomyx jubatus)*, 183, 229–30
Chelys, 41
chemical defences, 37, 84–95 (see poisons, venoms)
Chironex, 116–17, **117** (see sea-wasps)
chitin, 64–6, 80
Chlamyphorus, 70, **175**, 175–6 (see armadillos)
Choloepus, 189
Chorizops, **24** (see phragmosis)
chromatophores, 32
chrysalids, 47–52, 73, 137 (see pupae)
Chrysopa, 255 (see lace-wings)
Chrysopelea, 185
cicadas (Cicadidae), 134–5
Cichlidae, 208
ciguatera poisoning, 93
Cleopatra, 122, 125
clocks, biological, 26–7
Coatonachthodes, 224
cobras *(Naja)*, 121, 122, 125–7, 160, 172 (see hamadryad, ringhals); black-necked *(N. melanoleuca)*, 122; Egyptian *(N. haie)*, 122, 127; Indian *(N. naja)*, 122, 123, **160**
cockroaches (Blattidae), 85, 88, 89
cocoons, 47–9, 73
Coelenterata, 57–8, 63, 115–16, **117**, **118**
coffer-fishes *(Tetrosomus)*, 64, **65**
Colobopsis, **24**
colours, aposematic, 72, 83, 94, 95–6, 127, 128–33, 141; black, 131–2; cryptic, 28–32, 35, 37, 94, 128, 131, 133, 139, 157, 180, 222 (see camouflage); desert, 31; disruptive, 36–7; green, 30–1, 189
colour change, 31–2, 44, 54, 178
Columbus, C., 204
commensalism, 19–20, 54, 77, 214
conditioning, 148
cone-shells *(Conus)*, 119, **120**
C. geographicus, 119
Conidae, 119
corals, 57–8, 60, 116 (see Coelenterata)
coral-snakes, 123, 152–3

Corallium, 57 (see corals)
cornicles, 213
corncrakes *(Crex crex)*, 164
Cossyphus, 42–3, **43**
Cott, H. B., 94, 155
Cottus, 82 (see fishes, poisonous)
countershading, 29, 33, 34
cowries (Cypraeacea), 59
cowbirds *(Molothrus)*, 217–19
crabs, 20, 35, 54, 77, 133, 169, 172, 213, 214; *(Potamocarcinus)*, 172
crab-spiders (Thomisidae), 47
critical distance, 53, 181, 190, 193, 199
crocodiles (Crocodylidae), 68, 200; Nile *(Crocodylus niloticus)*, 227
Cronwright-Schreiner, S. G., 182
Crotalinae, 124
Crotalus, 124 (see rattlesnakes)
crown-of-thorns starfish *(Acanthaster planci)*, 57–8, 79–80
crypsis, 28–32, 33, 37, 94, 128, 131, 133, 139, 157, 180, 222
cryptozoa, 18, 25–7
Ctenophora, 32
cuckoo *(Cuculus canorus)*, 73
cuckoo-spit, 48
cuticle, arthropodan, 64–6
cuttlefish *(Sepia)*, 32, 63, 172–3, 178, 188
cycles, population, 227
Cyclosa, 53, 167 (see orb-web spiders)

Damsel-fishes (Pomacentridae), 214
darkling beetles (Tenebrionidae), 66, 84, **85**
Darwin, C., 14, 98, 103, 140, 151
Dasyuridae, 162
date-shells *(Lithophaga)*, 23
Dawkins, R., 208
death feigning, 163–4, 223
death's-head hawk-moth *(Acherontia atropos)*, 134, **134**
deer (Cervidae), 162–3, 191, 194, 195–6; red *(Cervus elaphus)*, 192, 195, **197**; roe *(Capreolus capreolus)*, **197**, 211; white-tailed *(Odocoileus virginianus)*, 222
deimatic (startling) behaviour, 158–63
Dendraspis, 123 (see mambas)
Dendrobates, 93
denticles, 81
Descartes, 105
desert coloration, 31 (see black coloration)
desert-beetles, 131–2, 151 (see Scarabaeidae, Tenebrionidae)
Didelphis, 163, **163**
dinosaurs, 67–9, 199–200

245

Index

Diploptera, 89 (*see* repugnatorial glands)
disguise, 40–55, 139, 141
Dispholidus, 125, 160
disruptive coloration, 36–7
dogs, 212 (*see* hunting dogs)
dog-whelks (*Nucella*), 60
dogfishes (Squaloidea), 80–1, 82
Dorippe, 51
Dorylenae, 67, 104
Draco, 164, 185, 187
dragons, flying (*Draco*), 164, 185, 187
dragonfish (*Trachinus* or *Pterois*), 81–2
dragon-lizards (*Goniocephalus*), 223
driver-ants (Dorylinae), 67, 104, 108
dromedary (*Camelus dromedarius*), 162
Dromia, 77, **213**
drones, 104
drone-flies (*Eristalis*), 140, 147
ducks (Anatidae), 94–5 (*see* mallard)
Dysdera, 46, **47**

Ears, 39, 179
earwigs (Dermaptera), 88
earthworms (Lumbricidae), 20, 21, 23, 26, 80, 85, 176, 203–4
echidna (*Echidna aculeata*), 72, 83, 228
echiuroid worms (*Urechis*), 19
Edmunds, M., 155
Edwardsia, 21
eels, electric (*Electrophorus electricus*), 97, 98–9
eggs, 36, 46, 90, 94, 165, 206
Elapidae, 122–3, 125 (*see* cobras, mamba)
electric fishes, 97–9, 100
Electrophorus, 97 (*see* eels, electric)
Eleodes, 84, **85** (*see* darkling beetles)
elephants, 12, 14, 122, 184, 196, 198, 199; African (*Loxodonta africana*), 196, 198; Indian (*Elephas maximus*), 196, 198
emigration, 227
Epinephalus, 178
Eristalis, 147
ermine, 30
ermine-moths (Arctiidae), 180
erratic (protean) behaviour, 157, 177–8, 201–2
escape reaction of scallops and queens, 61
escuerzo (*Ceratophrys cornuta*), 159
Euphractus, 70 (*see* armadillos)
Euproctis, 73
Eurythöe, 80 (*see* worms, Polychaeta)
excrement, resemblance to, 46–7, 139, 141 (*see* faeces, discharge)
Exocoetus, 186, **187** (*see* fishes, flying)

exoskeleton, arthropod, 64–6, 139
eyes, 18, 21, 25, 36, 52, 166
eye spots, 52, 161, 166
eye stripes, 36
eyed-hawk moth (*Smerinthus ocellatus*), 44–5

Faeces, discharge, 94–5 (*see* excrement, resemblance to)
false cocoons, 48
false-scorpions (Pseudoscorpiones), 113
fer-de-lance (*Bothrops atrox*), 125
fire-ants (*Solenopsis*), 102, 210–11
fireflies (*Photuris*), 155
fishes, 13–14, 33–4, 36, 44, 54, 80–3, 92–3, 204, 208, 214; Cichlidae, 208; coral, 128, 178; electric, 97–9, 100; flying (*Exocoetus*), 186, **187**; gurnard (*Dactylopterus*), 186, **187**; poisonous, 81–3, 92–3, 119, 159
fish-eagles, African (*Haliaeetus vocifer*), 227
flash coloration, 164, 178, 222
flatfishes, 54
flies, 133, 150–1 (*see* drone-flies)
flight, aerial, 185–8
flight distance, 181, 222
Flustra, 58
food hoarding, 231
Foraminifera, 58
formic acid, 84, 102
foxes (*Vulpes*), 71, 230
frogs, 92, 93, 129, 183; European (*Rana temporaria*) 159; flying (*Rhacophorus*), 186, **187**; horned (*Ceratophrys*), 159; poison-frogs (*Dendrobates*), 93
fugu (*Arothron*), 93
Fulgora, **50** (*see* alligator-bugs)
fungi, 215
Fungia, 57 (*see* corals)
funnel-web spiders (*Atrax*), 111, 112

Gaboon-viper (*Bitis gabonica*), 124 130
Galeodes, 66 (*see* camel-spiders)
galls, 215–17
gall-insects, 215–17
gall-wasps, 210
Gangara, 137
Gasteracantha, 167 (*see* orb-web spiders)
Gastropoda, 59
gazelles, Thomson's (*Gazella thomsoni*), 226, 229–30
geckos, flying (*Ptychozoon*), 35, 185
gene selection, 208
Geophilomorpha, 87

Index

ghost-crabs *(Ocypode)*, 35
Gila monster *(Heloderma suspectum)*, 121, 129, 139
gin-traps, 48, **74**, 74–5
giraffes *(Giraffa camelopardalis)*, 163, 194
Glomeris, 56 (*see* millipedes)
goat-antelope *(Oreamnos americanus)*, 192–3
goby fishes (Gobioidea), 214
godwits *(Limnosa)*, 178
Goniocephalus, 223
Gorgosaurus, 68
Graphosoma, 206
grasshoppers (Acridiidae, Tettigoniidae), 46, 89, 91, 92
grasshopper-mice *(Onychomys)*, 90
grass-snakes *(Natrix)*, 93–4, 161, 163–4
green coloration, 30–1
gribbles *(Limnoria)*, 21–3, **22**
grison *(Galictis vittatus)*, 96
grooming, mutual, 206
ground-beetles (Caribidae), 88, 90, 136; *(Pterostichus)*, 90
ground-squirrel, African, 220; *(Xerus)*, 175
group selection, 207–8
grunions *(Leuresthes tenuis)*, 204
gurnard, flying *(Dactylopterus volitans)*, 186, **187**

Habituation, 147–8, 176–7, 221
hadrosaurs (Hadrosauridae), 200
hairs, urticating, 73
hairstreak butterflies *(Thecla)*, 62
hamadryad *(Ophiophagus hannah)*, 11–12, 122–3
hares *(Lepus)*, 30
hawk-moths (Sphingidae), 44, 48–9, 137, 157, 159, 179; death's-head *(Acherontia atropos)*, 134, **134**; eyed-hawk *(Smerinthus ocellatus)*, 44 (*see* sphinx-moths)
hedgehogs *(Erinaceus)*, 71, 72, 126
Helaeus, 42
Heliconiidae, 140–1
Hemiptera, 130 (*see* bugs)
hermit-crabs (Anomura), 60, 213
hermit-ibis *(Geronticus)*, 94
Hermodice, 80 (*see* worms, Polychaeta)
herring *(Clupea harengus)*, 33
Hinton, H. E., 50
hippopotamus *(Hippopotamus amphibius)*, 162, 198–9
hissing, 133, 137, 139, 149, 158, 161–2 (*see* stridulation)
Holder, C., 98

honey-bees *(Apis mellifera)*, 102–4, 147, 209–10
honeydew, 212–13, 224–5
Horn, A., 198
horns, 162, 190–6
hummels, 195
hunting behaviour, 38–9, 50–1, 76, 147–8
hunting dogs *(Lycaon pictus)*, 183, 226, 228, 229
Huxley, J., 51
Huxley, T. H., 14
hyenas *(Hyaena)*, 145, **146**, 161, 205, 226, 229, 230
Hymenocera, 79
Hymenoptera, stings, 102–4

Ianthina, **34**, 59–60
Ichneumonidae, 149–50
Ictalurus, 98
Ictonyx, 96
Idolium, 158–9 (*see* mantids)
impala *(Aepyceros melampus)*, 211, 228
injury feigning, 165, 226
inquilines, 223–4
Ityraea, 205

Jackals *(Canis)*, 229
Jacobson's organ, 124–5
jararaca *(Bothrops neuweidii)*, 125
jays (Corvidae), 91, 92
jellyfishes, 32, 115–16, **117** (*see* Coelenterata)

Kin selection, 208
king cobras *(Ophiophagus hannah)*, 11–12, 122–3
Kinixys, 67
Kipling, R., 124, 165
Kirby, W., 140
kraits *(Bungarus)*, 123, 166

Lace-wings *(Chrysopa)*, 77, 224–5
Lampromyia, 75
lampshells (Brachiopoda), 62
langurs *(Semnopithecus)*, 212
Lachesis, 125
larvae, insect, 31, 46, 47–9, 73, 75, 77, 78, 86–7, 129, 130, 157, 161, 180, 205, 217–19
Latrodectus, 111, **111** (*see* malmignatte)
leaf-insects (Phasmidae), **42**, 157
leaves, resemblance to, 42, 43–5
lemmings *(Lemmus)*, 227, 229
Leodice, 203 (*see* palolo worms)
leopard *(Panthera pardus)*, 71, 206, 231
Leucorhampha, 157
Leuresthes, 204

247

Index

lichen, resemblance to, 45, 128
limbs, mammalian, 184
Limnitis, 51–2 (*see* butterflies)
Limnoria, 21, **22** (*see* gribbles)
limpets (Patellidae), 60
ling *(Molva molva),* 13–14
lion *(Panthera leo),* 37, 139, 180, 181, 190, 227, 230
lion-fish *(Pterois* and *Scorpaena),* 81–2
Lithophaga, 23
lizards, 67–8, 76, 168–9, 171 (*see* dragon lizards); Agamidae, 133; bearded *(Heloderma horridum),* 121, 129, 139; flying *(Draco),* 164, 185, 187; venomous *(Heloderma horridum* and *H. suspectum),* 120, 121, 129, 139
lobsters, spiny *(Palinurus),* 137
Lorenz, K., 208
Loxosceles, 111
lug-worms *(Arenicola),* 169
Lumbricus, 26 (*see* earthworms)
Lycaenidae, 136–7 (*see* butterflies)
lyre-snakes *(Trimorphodon),* 122
Lytta, 91 (*see* cantharidin)

Mackerel, common *(Scomber scombrus),* 33, 82
Macroclemys, 41
Malacochersus, **20**, 21
Malapturus, 97
malmignatte *(Latrodectus),* 111, **111**, 112
mallard *(Anas platyrhynchos),* 203
Mallophora, 146
Malpighi, M., 215
mamba *(Dendraspis angusticeps),* 122, 123
mantids (Mantidae), 46, 136, 158–9, 183, 222–3
matamata *(Chelys fimbriata),* 41
Meandrina, 57
meerkat *(Suricata suricatta),* 220, **221**
Melea, 214
Meloidae, 91, 92
Mephitis, **95**, 96
Mertens, R., 152
mertensian mimicry, 152–4
mice *(Mus),* 38, 172
millipedes (Diplopoda), 56, 65, 72, 87, 89, 108
mimicry, 140–56, 158, 223 (*see* ant mimicry)
minnow *(Phoxinus phoxinus),* 80
mites (Acari), 54
Mollusca, 34, 58–63, 169 (*see* sea-slugs, sea-snails)
Molothrus, 218 (*see* cowbirds)

mongooses *(Herpestes),* 12, 41, 126, 127; *(H. edwardsi),* 126 (*see* meerkat *(Suricata)*), 220, **221**
monkeys *(Colobus),* 128
monkeys' faces, resemblance to, 49–51, 136, 139, 157
Monocentrus, 64, **65**
moose *(Alces americana),* 194
Mormyridae, 97, 98, 99
moss-animalcules (Polyzoa, Ectoprocta), 58
moths, 41, 46, 130, 134–161; bag-worms (Psychidae), 76–7; blotched emerald *(Euchloris pustulata),* 54; death's-head hawk *(Acherontia atropos),* 134, **134**; ermine-moths (Arctiidae), 180; eyed-hawk *(Smerinthus ocellatus),* 44–5; *Gangara* (Hesperiidae), 137; gold-tail *(Euproctis chrysorrhoaea),* 73; owlet (Noctuidae), 179; red-underwing *(Catocala nupta),* 164; tiger-moths (Arctiidae), 137, 148, 180 (*see* hawk-moths, sphinx-moths)
mouflon *(Ovis orientalis),* 193
Müller, F., 148
müllerian mimicry, 148–9, 150–1, 154, 224
mullet *(Mullus),* 201
Mullus, 201
muntjac *(Muntiacus muntjak),* 193, **197**
mushroom-corals *(Fungia),* 57 (*see* corals)
musk-ox *(Ovibos moschatus),* 195, 201
mussels *(Mytilus),* 60–1, 180
mutillid wasps (Mutillidae), 133, 136
Mylogaulus, 192
Mydaus, 96
Myrmarachne, 142, **144** (*see* ant mimicry)
Myrmecobius, 76 (*see* banded ant-eater)
Myrmedonia, 223 (*see* ant mimicry)
Myrmeleon, 75 (*see* ant-lions)

Nabis, 143
Naja, 122 (*see* cobras)
narwhal *(Monodon monoceros),* 230–1
natural selection, 13, 39
nautiloids *(Nautilus),* 63
nematocysts, 57, 58, 63, 115–17, **118**, 130–1, 169
Neotragus, 184
Nephila, 225, 226
Neuroptera, 77, 224–5
nightjars *(Caprimulgus),* 165
Noctuidae, 179
nodosaurs (Nodosauridae), 68–9

Index

Nucella, 60
Nymphalidae, 51–2

Obliterative countershading, 29, 33, 34
octopods (Octopodida), 63, 119, 172–3
Odontosyllis, 204
Oecophylla, **144**, 145
oilbird *(Steatornis caripensis)*, 179
Onychophora, 84–5
Onychoteuthis, 188
Oniscus, 56 (see woodlice)
Ophiophagus, 122–3 (see hamadryad)
Ophrys, 217
opossums *(Didelphis)*, 163, **163**
orb-web spiders, 53, 73–4, 78, 158, 167–8, 225–6
organ-pipe corals *(Tubipora)*, 57
oropendulas *(Psarocolius)*, 217–19
Orsima, 143
oryx *(Oryx)*, 193, 194
ostriches *(Struthio camelus)*, 35, 46
otter, Central American *(Lutra annectens)*, 172
owls (Strigidae), 38–9; snowy *(Nyctea scandiaca)*, 229
owlet-moths (Noctuidae), 179
oysters *(Ostrea)*, 14

Palinurus, 137
palolo worms *(Leodice)*, 203–4
pancake-tortoise *(Malacochersus tornieri)*, **20**, 21
pangolins (Manidae), 64, **69**, 70, 76, 228
parasites, 26, 48, 86, 217–19, 227
partridge *(Perdix perdix)*, 165
pea-crabs *(Pinnotheres)*, 20
pearl-fishes *(Carapus)*, 174–5
peccaries (Tayassuinae), 96
pectines, **109**, 110, 208
pedicellariae, 79
peewit *(Vanellus vanellus)*, 178
Pentatomidae, 88, 130
Pepsis, 112–13
peregrine *(Falco peregrinus)*, 206
Petaurus, 185, **187**
phalangers, flying *(Petaurus)*, 185, **187**
Phasmidae, 42, 164
pheasants *(Phasianus colchicus)*, 202
Pheidole, 210 (see ants)
pheromones, 37, 86, 130, 208–11
Philornis, 218 (see botflies)
Pholas, 23
Pholidota, 228
Pholus, 157 (see hawk-moths)

photophores, 34
Photuris, 155
phragmosis, **24**
Phyllopteryx, 44
Physalia, 115–16
piddocks *(Pholas)*, 23
pigmentation, 13
pike *(Esox lucius)*, 80
pine-cone fishes *(Monocentrus)*, 64, **65**
Pirajno, A. D. di, 106
pit-vipers (Crotalinae), 123, 124–5 (see rattlesnakes); *Agkistrodon halys*, 123
Platymeris, 89 (see assassin-bugs)
platypus *(Ornithorhynchus anatinus)*, 83; venom, 81–3
Pliny, 73, 127, 215
plover, little ringed *(Charadrius dubius)*, 226
Poekilocerus, 92
poisons, 78–9, 80, 81, 87, 93, 101–27, 161, 217 (see chemical defences, venoms)
Polybius, 196, 198
Polyxenus, 72
poison-frogs *(Dendrobates)*, 93 (see frogs)
Porcellio, 56
porcupines (Erethizontidae, Hystricidae), 71–2, **138**, 138–9, 161
Portuguese man-of-war *(Physalia)*, 115–16
Potamocarcinus, 172
Poulton, E. B., 155
Probolodus, 225
Prociphilus, 224–5
pronghorn *(Antilocapra americana)*, 197
pronking, 181–2, 222
protean (erratic) behaviour, 157, 177–8, 180, 202
Proteles, 145, 161 (see aardwolf)
Psammophis, 122
Psammophylax, 122
Psarocolius, 218 (see oropendulas)
Pseudosphinx, 129 (see sphinx-moths)
Pseudolycaena, 166
Psychidae, 76–7
ptarmigans *(Lagopus)*, 29, **30**
Pterois, 82
pterodactyls, 186
Pterostichus, 90
Ptolemy III, 196
puff-adder *(Bitis arietans)*, 124, 129–30
puffer-fishes (Tetraodontidae), 159
pupae, 47–9, **49**, 50–2, 73, 74–5, 136–7, 151
python, African *(Python sebae)*, 128

Queens *(Chlamys)*, 61

249

Index

Rabbits *(Oryctolagus cuniculus)*, 30, 176
Radiolaria, 58
radula, 60, 79, 119
rapid peering, 51 (*see* birds, hunting methods)
rattlesnakes *(Crotalus)*, 124–5, 138, 161, 183
reactor glands, 89–90
Réaumur, R. A. F., 102
Redi, F., 105
Reduviidae, **114**, 115 (*see* assassin-bugs)
repugnatorial secretions, 37, 84–95, 132, 209, 210, 222 (*see* warning scents)
reverse mimicry, 52–3, 143–51, 166
Rhacophorus, 186
rhinoceroses (Rhinocerotidae), 69, 192
rhythms, diurnal, 26–7, 88
ringed-plovers *(Charadrius)*, 46, 226
ringhals *(Sepedon haemachetes)*, 122 (*see* cobras)
road-runners *(Geococcyx)*, 126
robber-flies *(Mallophora)*, 146–7
Rodentia, 38, 162, 172
Rotifera, 78
Royal antelope *(Neotragus pygmaeus)*, 184
rumination, 182–3
Ruskin, J., 232
Russell's viper *(Vipera russellii)*, 124

Safari-ants (Dorylinae), 67, 104, 108
Sagittarius, 184 (*see* secretary-bird)
salamanders, 93, 129, 183
samandarin, 93
Samurai, resemblance to, 51
sand-grouse (Pteroclidae), 202–3
sand-snakes, hissing *(Psammophis)*, 122
sand-wasps *(Ammophila)*, 176
sandhoppers (Talitridae), 177
Sargassum, 44
Saturniidae, 179
savanna animals, 184
saw-flies (Symphyta), 130, 205
saw-scaled viper *(Echis carinatus)*, 124, 137–8
scales, 64–8
scale-insects (Coccidae), 215
scallops *(Pecten)*, 61, 180
scaly lizards *(Sceloporus)*, 64
scarab-beetles (Scarabaeidae), 151
Sceloporus, 64
scents, 37, 138
sclerotin, 66, 86
scolopendras (Scolopendromorpha), 87, **113**, 113–15, 133, 166, 169–70 (*see* centipedes)

Scolosaurus, 68
Scorpaena, 81 (*see* scorpion-fishes)
scorpions, 104–10, **109**, 135
scorpion-fishes (Scorpaeonidae), 54, 81, 220; *(Scorpaena)*, 81
Scutigera, 170
Scytodes, 85
sea-anemones (Actiniaria), 21, 63, 130–1, 169, 214; *(Stoichachis)*, 214
sea-cucumbers (Holothuroidea), 174–5
sea-dragons *(Phyllopteryx)*, 44
sea-slugs (Nudibranchiata), 63, 117, 130–1, **131**, 169
sea-snails *(Ianthina)*, **34**, 59–60
sea-snakes (Hydrophidae), 123–4, 126
sea-urchins (Echinoidea), 78–9
sea-wasps *(Chironex)*, 116–17, **117**
searching image, 38, 147
seaweed, resemblance to, 44, 57
secretary-bird *(Sagittarius serpentarius)*, 126, 184
seeds, resemblance to, 42–3, **43**
Senecio, 92
Sergestes, 34
sharks (Selachii), 33, 80–1, 98
shelduck, Egyptian *(Alopochen aegyptiaca)*, 164; ruddy *(Tadorna ferruginea)*, 164
shells, 58–63
shipworms *(Teredo)*, 23
shrews (Soricidae), 120
shrimps (Natantia), 34, 214; *(Hymenocera)*, 79–80
silence, 38–9
silver-fishes (Lepismidae), 74
siphons, shipworm, 23
Siphonophora, 33
skaapstekers *(Tremorphodon)*, 122
skuas *(Stercorarius)*, 165–6, 171; Arctic *(S. parasiticus)*, 165; great *(S. skua)*, 171
skunks (Mephitinae), **95**, 95–6, 139, 161
sloths *(Choloepus)*, 189
slow-worm *(Anguis fragilis)*, 94
slugs (Gastropoda), 62–3, 85
snails (Gastropoda), 62, 176; banded *(Cepaea)*, 62 (*see* sea-snails)
snakes, 93–4, 120–7, 135, 139, 152–3, 163 (*see* cobras, coral-snakes, grass-snakes, hamadryad, mamba, rattlesnakes, sea-snakes, vipers)
snout-fishes (Mormyridae), 97, 98, 99
social behaviour, 206–7
solenodons *(Solenodon)*, 120
Solenopsis, 210 (*see* fire-ants)
Solifugae, 135 (*see* camel-spiders)

Index

Spalgis, **49**, 136, 139 (*see* monkeys' faces, resemblance to)
Spallanzani, L., 179
Spanish fly *(Lytta)*, 91 (*see* cantharadin)
spectacled caiman *(Caiman sclerops)*, **50**
speed mimicry, 151
Spence, W., 140
sperm-whale *(Physeter catodon)*, 188
spermatophores, 110
Sphingidae, 137, 157, 159, 179 (*see* hawk-moths)
sphinx-moths *(Pseudosphinx)*, 129; *(Leucorhampha)*, 157
spiders (Araneae), 24, 53–4, 56–7, 73–4, 77–8, 85, 110–13, 158, 167–8, 186, 225–6; ant mimics, 54, **142**, 142–4, 144, 223; black widow *(Latrodectus mactans)*, 111; brown recluse *(Loxoscles unicolor)*, 111; crab-spiders (Thomisidae), 47; funnel-web *(Atrax)*, 112; malmignatte *(Latrodectus tridecimguttatus)*, **111** (*see* webs, spider, orb-web spiders, wolf-spiders)
Spilogale, **95**, 96
spines, 71–3, 78–83
spiny ant-eater *(Echidna aculeata)*, 228 (*see* echidna)
springbuck *(Antidorcas marsupialis)*, 181–2
sponge-crabs *(Dromia)*, 77, **213**
squids (Decapoda), 33–4, 63, 172–3; flying *(Stenoteuthis, Onychoteuthis)*, 188
squirrels (Sciuridae), 145; flying *(Sciuropterus)*, 185, **187**
stabilimenta, 53, 167–8
stags, 190, 195 (*see* deer)
Staphylinidae, 223, 224
starfishes (Asteroidea), 57–8, 61, 64, 79–80, 180
star-gazers *(Uranoscopus)*, 99
starling *(Sturnus vulgaris)*, 206, 207
Steatornis, 179
Stenoteuthis, 188
stick-insects (Phasmidae), 41–2, **42**, 45–6, 88, 148
sticklebacks, three-spined *(Gasterosteus aculeatus)*, 80; ten-spined *(Pungitius pungitius)*, 80
stoat (ermine) *(Mustela erminea)*, 30
stonefishes *(Synancega)*, 82–3
stotting, 181, 222
Strabo, 126
stridulation, 133–7, 149, 170
Suricata, 220, **221**

surplus killing, 230–1
Swammerdam, J., 102
swiftlets *(Collocalia)*, 179
symbiosis, 214–17
Synancega, 82

Tanks, 11, 12, 40, 56, 85, 221
tails, autotomy, 171–2; resemblance to heads, 52–3, 166 (*see* reverse mimicry)
tarantulas (Mygalomorpha), 111, 112–13, 135; *(Lycosa tarantula)*, **111**, 112
tarantula-hawk *(Pepsis)*, 112–13
teeth, 102, 199
teledu *(Mydaus javanensis)*, 90
Telescopus, 121
Tenebrionidae, 66 (*see* darkling beetles)
tenrecs *(Centetes, Hemicentetes, Ericulus)*, 72
Teredo, 23 (*see* shipworms)
termites (Isoptera), 209, 223–4, 228–9; *(Nasutitermes)*, 86, 91, 104; *(Trinervitermes)*, 228–9; guests, 224
Tetragnatha, 167–8 (*see* orb-web spiders)
Tetrosomus, 64, **65**
Thecla, 166
Theophrastus, 215
thorns, resemblance to, 42–3
threat, 162
thrushes *(Turdus philomelos)*, 62
thylacine *(Thylacinus cynocephalus)*, 72
Tinbergen, N., 208
tiger *(Panthera tigris)*, 71, 139
tiger-moths (Arctiidae), 137, 148, 180
toads *(Bufo)*, 76, 93, 159
Tolypeutis, 70
Tonica, 48–9
Topsell, E., 120
Tornier's tortoise *(Malacochersus tornieri)*, **20**, 21
torpedoes *(Torpedo)*, 97, 99
tortoises (Testudinidae), 12–13, 20–1, 66–7, 94
Trachinus, 81
transparency, 32–3
tree-porcupines *(Erethizon)*, 71–2
tree-shrews (Tupaiidae), 145
tree-snakes *(Chrysopelea ornata)*, 185; *(Leptophis)*, 45
Triatoma, **114**, 155 (*see* assassin-bugs)
Triceratops, **67**, 68
Trimorphodon, 122
Trinervitermes, 228–9
triton-shell *(Charonia)*, 79
trunk-fishes *(Tetrosomus)*, 64, **65**

251

Index

Tubipora, 57
turacos *(Touraco)*, 31
tympanic organs, 179
Tyrannosaurus, 68

Uloborus, 167
ultra-sound, 38–9, 137, 178–9
uric acid, 100
urine, 94

Vanellus, 178
venoms, 102–27, 161 (*see* poisons);
  elapid, 125–6; fish, 54, 81–3, 220;
  mammalian, 81–3, 120; scorpion,
  104–10; starfish, 78–9; viperine, 125
Vermileo, 75
vervet monkey *(Cercopithecus aethiops)*, 212
vicuña *(Lama vicugna)*, 162
vinegaroon *(Mastigoproctus giganteus)*, 88 (*see* whip-scorpions)
vipers (Viperidae), 121, 124, 125, 126
Viperinae, 124
Voigt, W., 155
vomit, 94–5

Wallace, A. R., 14, 129, 140
warning (aposematic) coloration, 72, 83, 94, 95–6, 127, 128–33, 141, 148, 152–4, 173, 205–6, 217; scents, 138, 139; sounds, 72, 133–9, 153, 162, 178 (*see* hissing, stridulation)
wart-hog *(Phacochoerus aethiopicus)*, 190
Wasmann, E., 145, 223, 224
wasmannian mimicry, 145, 223–4
wasps, 102, 104, 112–13, 129, 132, 133, 135, 209, 211–12
water drops, resemblance to, 48
water-beetles (Dytiscidae), 88, 151

water-deer, Chinese *(Hydropotes inermis)*, 191, **197**
water-fleas *(Daphnia)*, 202
water-mites (Hydracarina), 148
water-mongoose *(Atilax paludinosus)*, 41
waxwings, cedar *(Bombycilla cedrorum)*, 206
webs, spider, 53–4, 112, 167–8
weevers *(Trachinus)*, 81–2
weevils (Curculionidae), 54–5, 77
whales (Cetacea), 188
whip-scorpions (Thelyphonida), 87–8, 113
whirligig-beetles (Gyrinidae), 88
White, G., 165
white-flies (Aleyrodidae), 74
Wickler, W., 155
wildebeest *(Connochaetes taurinus)*, 205, 227, 230
wings, insect, 186
wolf-spiders (Lycosidae), 112
wood-ants *(Formica)*, 84, 102
wood-boring beetles, 24
woodlice (Oniscoidea), 26–7, 56–7, **57**, 65
worms (Lumbricidae), 21, 23, 26, 176, 203–4; (Polychaeta), 23, 80, 169, 176–7, 203–4 (*see* palolo worms)
worm-snakes (Typhlopidae), 76
Wynne-Edwards, V. C., 207

Xenarchus, 135
Xenarthra, 228
Xerus, 175

Zebras *(Dolichohippus)*, 35–6, 190
zone-tailed hawk *(Buteo albonotatus)*, 154–5
zorilla *(Ictonyx striatus)*, 96, 161